Thomas Rudolphus Dallmeyer

Telephotography

An elementary treatise on the construction and application of the telephotographic

lens

Thomas Rudolphus Dallmeyer

Telephotography

An elementary treatise on the construction and application of the telephotographic lens

ISBN/EAN: 9783337275938

Printed in Europe, USA, Canada, Australia, Japan

Cover: Foto ©berggeist007 / pixelio.de

More available books at **www.hansebooks.com**

TELEPHOTOGRAPHY

AN ELEMENTARY TREATISE

ON THE

CONSTRUCTION AND APPLICATION

OF

THE TELEPHOTOGRAPHIC LENS

BY

THOMAS R. DALLMEYER, F.R.A.S.

VICE-PRESIDENT OF THE ROYAL PHOTOGRAPHIC SOCIETY

WITH TWENTY-SIX PLATES AND SIXTY-SIX DIAGRAMS

LONDON
WILLIAM HEINEMANN
1899

This Edition enjoys Copyright in all Countries Signatory to the Berne Treaty, and is not to be imported into the United States of America.

All rights including translation reserved

Dedicated

TO

THE MEMORY OF

MY FATHER

JOHN HENRY DALLMEYER

B. 1830—D. 1883

FAMED FOR HIS WORK

HIMSELF BELOVED

CONTENTS

CHAP.		PAGE
	FRONTISPIECE	
	PREFACE AND HISTORICAL NOTES	xi
I.	PROPERTIES OF LIGHT	1
II.	THE FORMATION OF IMAGES BY THE "PINHOLE CAMERA," AND ITS PERSPECTIVE DRAWING	5
III.	THE FORMATION OF IMAGES BY POSITIVE LENSES	17
IV.	THE FORMATION OF IMAGES BY NEGATIVE LENSES	41
V.	THE FORMATION OF ENLARGED IMAGES:	
	PART I. BY TWO POSITIVE LENS-SYSTEMS	51
	PART II. BY A POSITIVE SYSTEM AND A NEGATIVE SYSTEM COMBINED, OR THE TELEPHOTOGRAPHIC LENS	53
VI.	THE USE AND EFFECTS OF THE DIAPHRAGM, AND THE IMPROVED PERSPECTIVE RENDERING BY THE TELEPHOTOGRAPHIC LENS	80
VII	PRACTICAL APPLICATIONS OF THE TELEPHOTOGRAPHIC LENS	114
VIII	WORKING DATA	137
	ABRIDGED FORMULÆ FOR REFERENCE	142
	BIBLIOGRAPHY	147

PREFACE AND HISTORICAL NOTES

This treatise is addressed to those who practise photography either for pictorial or scientific ends.

The late Michael Faraday once remarked: "Lectures which really teach will never be popular; lectures which are popular will never teach." All writers must experience the same inherent difficulty of treating scientific matter in any but an academic style. The author has endeavoured to present the subject of Telephotography in a manner which presupposes only the very slightest acquaintance with the science of Optics, explaining fully only those few properties or functions of lenses, which are necessary to enable the photographer to understand the action of the Telephotographic lens, and to comprehend the wide possibilities of its applications.

The aim of the present work, in short, has been to call attention to the *scale* in which objects are reproduced in the image by ordinary photographic lenses, and to show how this image may be subjected to direct enlargement or magnification before it is received on the photographic plate.

This method was adopted by the author in his original contributions to the subject. It is perhaps less classical than treating the instrument as a complete optical system, but has a more practical bearing upon its use; and as might be expected, is more readily grasped by those who are acquainted with the action of ordinary photographic lenses: both methods are, however, included. It is

PREFACE AND HISTORICAL NOTES

hoped that frequent diagrammatic representation and occasional repetition may be of assistance to the reader. The application of the few formulæ given in the work only involve a knowledge of arithmetic.*

The literal meaning of the term "Telephotography" does not convey the full significance of the applications of the Telephotographic lens. The value of the instrument is as great, if not greater, in photographing near as well as distant objects. It will be found to possess invaluable properties wherever lenses of great focal length are required to produce large images on the one hand, and for rendering improved perspective drawing in any given scale of near objects on the other. This latter effect is brought about by the fact that a greater distance intervenes between a Telephotographic lens and a comparatively near object (as in portraiture) than that required when using a lens of ordinary construction of the same focal length. This property is of great value to the artist when the object he desires to photograph has any "depth of field" enabling him to avoid the apparent exaggerated perspective so frequently met with in ordinary photographs.

The principle upon which the Telephotographic lens is constructed has been applied to the astronomical telescope for nearly seventy years. So long ago as 1834, Peter Barlow, in a communication to the Royal Society, dwelt on the advantages that might accrue from employing his *negative* lens "in day telescopes" as well as in astronomical telescopes; "for by giving an adjustment to the lengthening lens, the power may be changed in any proportion, without even removing the eye or losing sight of the object. I have no doubt these and other applications of the lengthening lens will be made." In these last few words Barlow foreshadowed the construction of the Telephotographic lens.

* The notes to Chapters II., III., IV., V., and VI. may be omitted on first reading.

PREFACE AND HISTORICAL NOTES

Dr. Von Rohr, a contemporary of the present writer, has been at great pains to trace the application of Barlow's lens in the construction of *photographic* instruments; and it appears that in 1851 Porro, an Italian engineer, utilised a " Barlow" lens for photographing an eclipse of the sun which took place on July 18 in that year. Since that date, the employment of a negative lens in astro-physical work has been adopted in a few isolated instances, notably by Dr. H. Schroeder, and these instruments are reported to have been directed to very distant terrestrial objects for the purpose of photographing them. Ordinary astronomical and terrestrial telescopes of various kinds have been utilised for the same purpose from time to time, and in the year 1873 the late Mr. J. Traill Taylor, Editor of the *British Journal of Photography*, called attention to the use of the Galilean Telescope, or ordinary opera-glass, for the purpose of producing a large direct image. This application of the Galilean telescope is identical with the employment of the " Barlow" lens for direct enlargement of the image, and as the instrument was not designed or corrected for photographic purposes, the reference to the opera-glass, although occurring in photographic literature, was unnoticed elsewhere.

In February 1890, Steinheil constructed a special photographic instrument upon this principle for the German " Reichs-marineamt," but the fact was not published.

In the autumn of 1891, A. Duboscq, Dr. A. Meithe and the author almost simultaneously applied for patents concerning Telephotographic instruments; Duboscq in France on August 7, Dr. Miethe in Germany on October 18, and the author in England on October 2. Duboscq's work was not known till a reference was made to it by A. Sorets in 1893.

A controversy between the author and Dr. Miethe as to priority took place in the *British Journal of Photography*, in which the author acknowledged Miethe's independence. At this date there had been no

PREFACE AND HISTORICAL NOTES

previous publication of instruments designed for the use of photographers.

The author was the first to exhibit instruments so designed at the Camera Club, London, and to explain the theory of construction and working. This is acknowledged by Dr. Von Rohr (of the firm of Zeiss), who points out the widespread effect of this exposition.

The inherent defect of distortion of the image in the first Telephotographic constructions led the author to introduce the plan of converting ordinary non-distorting photographic lenses into Telephotographic systems, without interfering with their ordinary use, and this form of combination is the one which is now chiefly adopted by his contemporaries. Full reference is made to particular constructions.

The author desires to record the fact that his attention was first directed to the subject of Telephotography by his friend Dr. P. H. Emerson, who urged upon him the necessity of a photographic instrument to enable the naturalist to record incidents that were then only possible by telescopic observation. Emerson's original work in advancing the pictorial side of photography is now history, and it was his indication of certain drawbacks in photographic methods, as a means of pictorial representation, particularly the inadequate rendering of objects as *seen* by the normal eye, that led the author to endeavour to overcome them by improved optical means.

M. Boissonnas, of Geneva, the Earl of Crawford, and Mr. Hodinott, of the Camera Club, London, were the first to test and prove the value of the instrument in distant mountain scenery, and Dr. Emerson (in 1892) and later Mr. J. S. Bergheim to exhibit results showing improved perspective in portraiture. The author's thanks are due to these gentlemen and many others for their kindness in offering examples to illustrate this little work. Mr. R. B. Lodge and Messrs. Kearton have exemplified its value to the naturalist, Mr. E. Marriage and Mr. Cruickshank to the architect, while Dr. Victor Corbould and the late Dr. Fallows have

PREFACE AND HISTORICAL NOTES

demonstrated its utility in surgical and medical records; some of their splendid work would have been here reproduced, except for reasons that will be obvious when it is remembered that this treatise is intended for perusal by the lay public. A few studies of the eye are therefore substituted. The Astronomer-Royal kindly lends two very interesting examples of solar photography, while Naval and Military records and possibilities are illustrated through the courtesy of representatives of the Japanese and Italian Governments.

It may be mentioned that with the exception of one or two articles on the practical applications of the lens by Mr. Lodge, Mr. Marriage and Dr. Spitta, the subject has not been treated by any other English writer.

The author's papers are scattered—some being out of print. He therefore hopes that this treatise may find acceptance, and a place in the literature of photographic optics.

ROYAL SOCIETIES CLUB,
    ST. JAMES'S STREET, W.
      September 1899.

TELEPHOTOGRAPHY

CHAPTER I

PROPERTIES OF LIGHT

Light consists of vibrations of a highly elastic solid medium termed ether, which pervades all space. These vibrations are conveyed to the brain by means of our eyes, and produce the sensation of sight.

All visible objects surrounding us are sources of light. These sources are either *self-luminous bodies*, such as the sun, fixed stars, electric light, and bodies in a state of combustion; or *illuminated bodies*, as the moon, planets, or any body that can be seen by light borrowed from a self-luminous body.

For our present purpose, it is not necessary to distinguish these two classes of objects; every point in every object is a source of light.

Medium.—Any space through which light can pass is called a *medium;* a vacuum, air, gases, water, glass, &c., are media. Bodies through which light can pass are termed transparent, while those through which it cannot pass are termed opaque.

Light travels in straight lines with uniform velocity in any homogeneous medium. In the practice of photography, particularly in Telephotography, we are occasionally reminded of the converse of this law. In observing a distant object across an open landscape, or a ship

TELEPHOTOGRAPHY

at sea, the object under observation sometimes appears to be disturbed and unsteady, owing to a kind of "boiling" appearance of the atmosphere, which is not homogeneous for the time being. When the atmosphere is in this condition, it is hopeless to expect to obtain well-defined photographic images. The observations of astronomers in this country are frequently disturbed by the lack of homogeneity in the atmosphere, even on apparently the clearest nights.

Light consists of separable and independent Parts.—If we place an opaque object in the light proceeding from a luminous body, a portion of the light is intercepted, but the rest of the light proceeds to illuminate surfaces upon which it falls, neither adding to nor diminishing the illumination. Again, light from two independent sources may travel along the same path without interference. Hence light is capable of quantitative measurement; that is to say, we can compare the intensity of two given sources of light, or in general measure the intensity of light in terms of some fixed standard. In photographic practice, we measure the intensity of light by an instrument known as an "actinometer," the action of which is based upon the period of time taken by the light to discolour a photographically active material to a given shade; the greater the intensity of the light (photographically), the shorter the period of time taken to discolour the sensitive material.

Absorption and Transmission of Light.—When light travels through any homogeneous medium, part is absorbed and only part transmitted. We shall neglect absorption entirely and presume that all media are perfectly transparent.

Reflexion of Light on passing from one Medium to another.—When light passes from one homogeneous medium to another, part is reflected at the surface of the second medium, and part transmitted. In the construction of photographic instruments, we consider the whole of the light to be transmitted. In our study of the theory of the Telephotographic lens, we shall not then take these factors into consideration; but, in practice, it must be borne in mind that the absorption of light by thick lenses, and the loss of light by reflexions are not in reality

PROPERTIES OF LIGHT

quite negligible quantities. Greater optical perfection in an instrument is frequently attained, however, by both these means as exemplified in some of the most recent advances in photographic lens construction.

A Ray of Light.—For theoretical considerations, it is often convenient to consider a portion of light travelling along some particular line in a medium, quite apart from the remainder. We may think of it as an indefinitely attenuated slender cone, or, as older writers put it, the least portion of light we can conceive to exist independently. We term such a portion of light a Ray of Light.

A Pencil of Rays is a collection of rays which never deviate far from some central fixed ray, which is called the *axis* of the pencil. If the

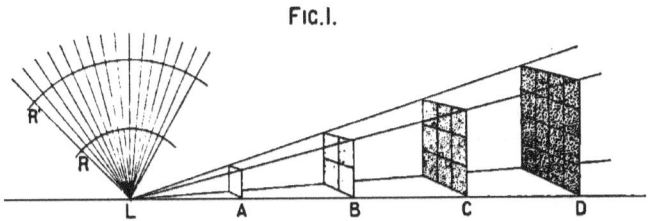

FIG.I.

pencil proceeds towards a point, it is termed a *convergent* pencil; if, on the other hand, it proceeds from a point, it is termed a *divergent* pencil.

Focus.—If a pencil of rays meets in a point, that point is called the focus. A focus is not a measurement, but a position.

Parallel Rays.—The form of a pencil of rays is considered as that of a right cone. The limiting form of a cone when the vertical angle is indefinitely small is a cylinder. Such a cylindrical pencil of rays is termed parallel.

The intensity of light at different distances from a luminous point is inversely as the square of the distance.

As we know that light travels in straight lines in all directions, it is easy to find the alteration in the intensity of the illumination* of a surface by changing its distance from the source of light.

* It is thought unnecessary to introduce "units" into this popular explanation.

TELEPHOTOGRAPHY

If L be a luminous point, and we imagine concentric spheres with radii R R', &c., described about it, the total quantity of light received by the surface of the sphere of radius R L is the same as that which would reach the surface of the sphere of radius R' L. Now, as the surfaces of the spheres are in direct ratio of the square of their radii, the quantity of light which falls on a given extent of surface must be inversely in the same ratio.

In the same manner, if we place a *small* plane surface at A at a given distance from L, the whole quantity of light received at A would be distributed over B, C, and D placed at distances 2, 3, and 4 times the distance of A from the light. The areas of B, C, and D are 4, 9, and 16 times as great as A, or the intensity of the light on the unit of surface is inversely as the square of the distance from the source of illumination. This law is termed the "law of inverse squares," and it has a very important bearing upon our present subject.

The illumination of the areas we have just discussed is taken as perpendicular to the incident light.

The illumination of an area inclined to the incident light is found by multiplying the former by the cosine of the angle of incidence. This law, together with the law of inverse squares, enables us to measure the falling off of the illumination from the centre towards the edge of a photographic plate, but need not be dwelt upon here.

CHAPTER II

THE FORMATION OF IMAGES BY THE "PINHOLE CAMERA," AND ITS PERSPECTIVE DRAWING

THE simplest of all devices for forming an image is what is termed a "Pinhole Camera."

If we make a minute hole in a thin sheet of card or metal, and place this at one end of a [rectangular] light-tight chamber or camera, and place a screen or sensitive photographic plate at the opposite end,

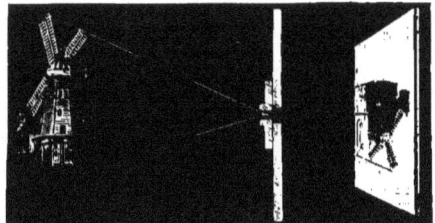

it is easy to see that any luminous or illuminated object lying in front of this camera will form an image upon the screen.

Every luminous point that contributes to make up the entire object emits, as we have seen, luminous rays in all directions, but only one very small pencil of light can pass from each separate point through the tiny hole. Each minute pencil passes through this pinhole in straight lines, and passes on until it is intercepted by the screen, where

TELEPHOTOGRAPHY

it is received as a tiny dot of light, determining its image. Thus the entire object forms a complete image upon the screen or sensitive plate, as illustrated in the figure. (See note at the end of chapter.)

Successful photographs have been obtained by this means in cases where extremely fine definition is not of prime importance, nor the requisite time of exposure (a few minutes in good light) likely to emphasise any defect in definition by movement of the object.

The enthusiastic photographer, having inadvertently left his lens

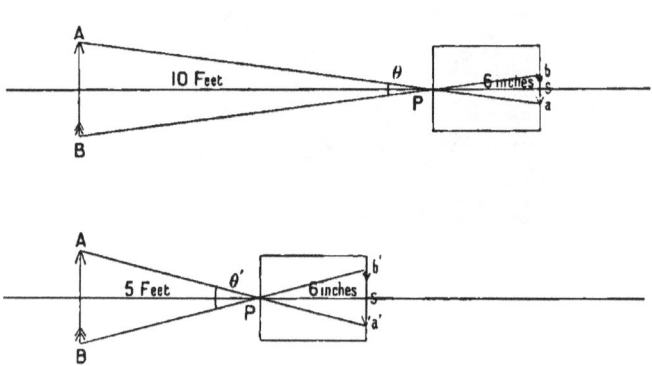

FIG. 3.

at home, has before to-day fixed his visiting card over the flange in his camera, pierced a hole in the pasteboard with a pin, and by this means secured a photograph that otherwise might never have been his, or fallen to his lot again.

The "Pinhole Camera" gives us a clear insight into the formation of images of different sizes, and will help us to understand similar effects when we come to examine the capabilities of lenses in this respect.

First let us direct the camera towards an object situated at a given distance from the pinhole, the distance between the pinhole and the screen being also known.

FORMATION OF IMAGES

Let us suppose the object A B (Fig. 3) to be 10 ft. distant from the pinhole P, and that P is 6 inches from the camera screen S; it is evident that the length of the image $a\,b$ formed upon the screen bears the same proportion to the length of the object A B as the distance P S (6 inches) is to P O (10 ft.), or the image is one-twentieth of the size of the object.*

Now let us bring the camera nearer to the object, as shown in the same figure, until O P is only 5 ft., but O S remaining 6 inches. The image now formed upon the screen is affected in a similar manner to our impression of the appearance of the object. As we approach an object it appears to be larger, or the object is said to be viewed under a greater angle, and it now subtends a greater angle at the pinhole

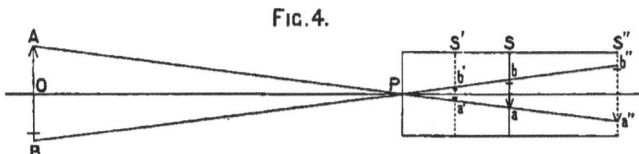

Fig. 4.

than it did in its first position. The rays passing from A and B through the pinhole cross at a greater angle. These and all rays forming the image are intercepted by the screen, and the length of the image is again determined by the relation of the distances P O (5 ft.) and P S (6 inches), A B and $a\,b$ being in the same proportion, or as 10 : 1.

Thus we see that when the camera is one half the distance from the object, the image is double the size (linear); and in general the nearer we approach the object, the greater becomes the size of the image. By similar reasoning the converse may be taken for granted.

Let us now consider the case of the object and pinhole of the camera at a fixed distance apart, but with the screen of the camera made to occupy different positions.

* A P O and a P S are similar triangles, and if we know the relation existing between the measurement of any two similar sides, O P : P S :: 20 : 1, we know that any other pair of similar sides O A, a S, bear the same relation to one another. Thus in the similar triangles A P B, a P b, A B : $a\,b$ :: 20 : 1.

TELEPHOTOGRAPHY

Let the object A B (Fig. 4) be again 10 ft. distant from P, and the screen S 6 inches from P, intercepting the rays of light from the object A B which pass through P, receiving the image at $a\,b$. In this position we have seen that the respective lengths of A B and $a\,b$ are as 20 : 1, or as their distances from P.

Let us now move the screen nearer to P, as at P S', say 3 inches only from the pinhole. We observe that the luminous rays from A B are intercepted sooner, and form an image at $a'\,b'$. The relative lengths of A B and $a'\,b'$ are now as 10 ft. to 3 inches, or as 40 : 1. Similarly, it is easy to see that if the screen be removed to S'', at a distance of 12 inches from P, the image will now be intercepted at $a''\,b''$, and that the relative

FIG. 5.

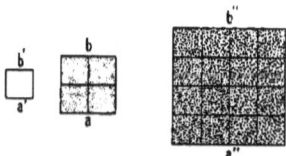

lengths of A B and $a''\,b''$ are now only as 10 : 1 ; and, in general, that the greater the distance between pinhole and screen, the larger will the image become.

It will be seen later that when we employ an ordinary photographic lens, the only way to increase the size of the image upon the screen is by approaching the object, as we did with the pinhole camera in Fig. 3 ; but when we employ a Telephotographic lens, we can place the screen in any position we like, as in Fig. 4, and obtain different sizes of images from a fixed standpoint. So that the pinhole camera with the screen in a fixed position roughly defines the limited use of an ordinary photographic lens as regards its power of producing images of different size of a given object ; whereas the pinhole camera with a movable screen indicates the far wider possiblities of the Telephotographic lens in this respect.

In Fig. 4 reference has been made to the *lengths* of the images $a'\,b'$,

FORMATION OF IMAGES

ab, $a''b''$ corresponding to A B at different distances of the screen from P, or different camera-extensions as they are termed. These lengths were found to be 1, 2, and 4 respectively. Let us suppose the object to be a square; then the image at $a'b'$ will be one unit square, the image at ab, two units square, and at $a''b''$ four units square.

It is evident that the quantity of light received at $a'b'$ (Fig. 5) is spread over four times the same area at ab, and sixteen times the same area at $a''b''$. The correct relative exposures to give to a photographic plate at these distances from the pinhole P are proportional to the areas in the figure, provided the aperture at P remains the same. We see thus early in our investigations the importance of the "law of inverse squares" in photographic practice: ab is twice the distance of $a'b'$ from

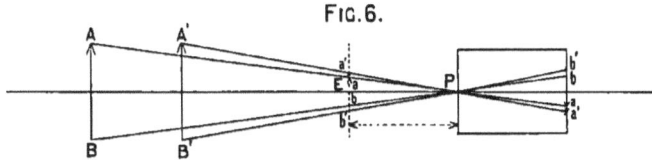

FIG. 6.

the source of light P, and receives one-fourth of the light; $a''b''$ is four times the distance of $a'b'$ from the same source, and therefore receives but one-sixteenth of the light.

It must be remembered that when we speak of the relative sizes of object and image, or of two images, we always refer to *linear* measurements. For convenience we shall frequently speak of "magnification" in this treatise, meaning invariably *linear magnification*. If we refer $a''b''$ to $a'b'$, for example, we say that the *magnification* of $a''b''$ is four times; we also speak of either reduction or magnification (in their ordinary sense) as "magnification." Thus in Fig. 4 the "magnification" of A B at $a'b'$, ab, $a''b''$ is $\frac{1}{10}$, $\frac{2}{10}$, and $\frac{4}{10}$ respectively.

The pinhole camera will assist us materially in a preliminary inquiry into the perspective drawing given by lenses. We shall, however, go into the matter more fully in due course.

Let us suppose we have two objects A B, A' B', (Fig. 6), of the same

TELEPHOTOGRAPHY

height at a given distance apart, and that the camera is made to approach A B, until the image *a b* is conveniently included upon the screen or plate; it will be obvious that *a' b'* the image of A' B' is also included upon the plate, and is smaller than *a b*.

If the eye be placed at P, A' B' subtends a greater angle than A B, and hence A' B' appears larger than A B.

In order to see the entire image *a b a' b'* in true perspective subsequently, it is necessary to consider the image projected forwards through the pinhole towards the object, until, when viewed from P, this image

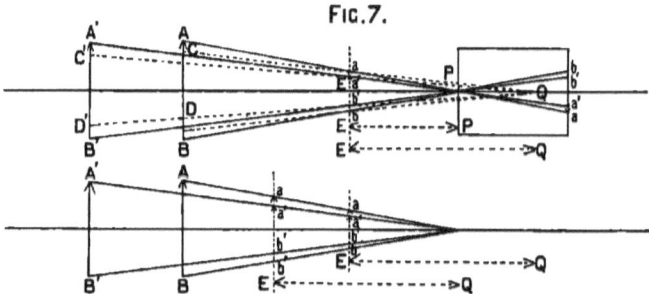

FIG. 7.

exactly overlaps, or coincides with its further projection on to the object itself. P, or the pinhole, is termed the "Entrance Pupil" (Abbé) of this image-forming appliance.*

The distance between this "pupil" and the image—when it is made to occupy a position in which its projection towards the object coincides with the object—determines the proper viewing distance for the image or photograph. We may say in passing that this consideration has a very important bearing on the perspective given by the Telephotographic lens.

It is evident from Fig. 6 that it is immaterial how near, or how

---

* There are two "Pupils" in every lens-system, termed the "Entrance" and "Exit" Pupils of the system, which are the centres of perspective for object and image respectively. In the case of pinhole projection, they coincide in the pinhole itself. (See Chapter VI., and Notes to same.)

FORMATION OF IMAGES

distant, the screen or plate is placed to the pinhole P in order to produce theoretically perfect perspective, because a position can always be *found* in front of the "Entrance Pupil" P, where this image, when projected towards the object, coincides with it exactly. (In the case under consideration the correct viewing distance is identical with the measure of the distance between the pinhole and the screen. It is, however, a less comprehensive method of examining the perspective drawing of photographic instruments in general.)

The theoretically perfect perspective above referred to may, and frequently does, impose conditions that are unsuitable, and often impossible, to observe if we possess normal powers of vision. We cannot, without effort or discomfort, look at an object that is nearer than about 10 inches. The result of this is that we usually view small photographs from too great a distance, and accordingly the image appears crowded into a smaller space than it ought to be.

In Fig. 7 the image $a\,a'\,b'\,b$ appears in true perspective from P; if, however, P E is not a convenient distance for normal vision, we should in reality view it from the same position Q. From this point Q the conditions for true perspective do not obtain, and the projection of the image towards the object does not coincide with it, but falls within it as at C D, C' D', giving an appearance of crowding.

It is evident that if a distance Q E is necessary to view the picture (image) in comfort, the only way to maintain correct perspective will be to project $a\,a'\,b'\,b$ towards the object until it would occupy the position at E', where E' P = Q E in the top drawing; in other words, the image must be enlarged. It may be stated here that enlargements are usually more satisfactory in perspective than small photographs, because the tendency is to view the latter from a greater distance than the theoretical conditions for true perspective allow. On the other hand, if we look at a photograph at a normal distance of vision, and this happens to be within the correct distance for true perspective, we do not feel that the perspective drawing given by the instrument is as unsatisfactory as in the former case; although if carried to an exaggerated degree, we become sensible that objects in the receding

TELEPHOTOGRAPHY

planes of the picture appear to be rendered too nearly upon one plane, or, to use the common phrase, "flat."*

In ordinary photographic practice, the camera is usually brought very near to the chief object of interest, so as to give it prominence by obtaining a sufficiently large image. When this takes place, the receding planes of the picture are dwarfed, while the perspective is unsatisfactory; for we do not so view it that its projection from the point of sight coincides with the objects in space. By the use of the pinhole camera it can readily be seen how this may be overcome.

In Fig. 8, where P is at a given distance from the object A B in order that ab may be the required size upon the screen S S, at a fixed

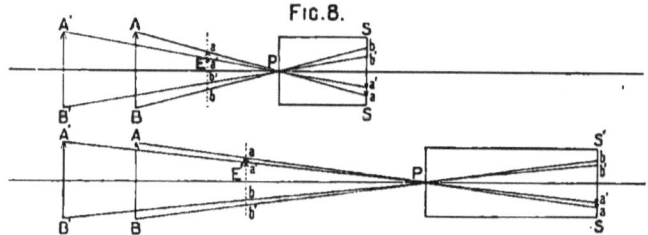

distance from P, the objects A B, A' B' are rendered in a certain proportion $ab : a'b'$; $a'b'$ being considerably smaller than ab. P E is here the correct but inconvenient viewing distance.

If we now remove the camera to a greater distance from A B, the image diminishes in size, and becomes too small for our purpose. But if the screen S S be removed from P, so that the image is received upon S' S' in the same size as in the former case, we shall find the image $a'b'$ of A' B' is now greater than before, or that the proportion $ab : a'b'$ is different in this case. In general, the perspective will be much more satisfactory if the nearest object of interest is not too close to the camera. As already stated, the conditions in the upper drawing in Fig. 8 are very approximately those of an ordinary "positive" lens;

* The reader is referred to Dr. P. H. Emerson's "Naturalistic Photography" (Third Edition, Messrs. Dawbarn Ward) for a fuller study of the perspective rendering of lenses.

FORMATION OF IMAGES

whereas in the lower drawing we exhibit the possibilities of the Telephotographic lens.

It is generally held that we ought to view a photograph from a distance equal to the focal length of the lens with which it was taken. This is very approximately true for all lenses irrespective of their construction, when the object is so distant that all rays meeting the lens may be considered parallel; but *only* under these conditions.

To see the image in true perspective under all conditions we must view it projected towards the object from the "Entrance Pupil" (of the particular optical system with which it was produced), at the *position where its projection towards the object will exactly coincide with the object.*

There is, then, a definite and correct distance at which every photograph should be viewed, but if we ask ourselves how often we conform to the correct conditions, we shall have to confess that it is very seldom indeed. It is for this reason that many artists are so severe upon the rendering of perspective by photographic instruments. It is not difficult to see how this comes about. If we look at a painting, a black and white drawing, or a photograph taken by ordinary means, whatever its size may be, we involuntarily take up a position at a distance equal to two, three, or possibly more times the length of its longest side. This standpoint suits the artist's perspective, and it *looks right*, because he has considered all this in creating his impression; but the poor photograph is terribly handicapped, because we *ought* to inspect it from a distance equal approximately to the focal length of the lens with which it was produced (generally about the longest side of the print) and when we do not, it looks wrong!

The Telephotographic lens, as will be shown anon, acts as a lens of very considerable focal length, and with its aid photographs are produced that *ought* to be viewed from a distance equal to three or four times the longer side of the print. We thus conform to our conditions of finding the correct position for true perspective involuntarily, and are not troubled with that apparently false rendering of perspective given by ordinary lenses.

TELEPHOTOGRAPHY

NOTE.

By ordinary reasoning, it might be presumed that the smaller the "pinhole" the finer the definition. This, however, is not the case, as light in passing through the small hole is "diffracted." This "diffraction" interferes with the definition of the image, and for a given distance between screen and pinhole the size of the hole may be either too large or too small to give the best result.

Diffraction is due to the retardation of rays proceeding from the margin of the pinhole as compared with rays proceeding from the centre; to obtain the maximum concentration of light at the focus, the retardation should be half a wave length $\frac{\lambda}{2}$.* Lord Rayleigh has shown at any rate that the *limit* of retardation may be taken as $\frac{\lambda}{4}$.

Taking $\frac{\lambda}{2}$ as the retardation, r as the radius of the pinhole, and d its distance from the plate P, it is easily seen that

FIG. 9.

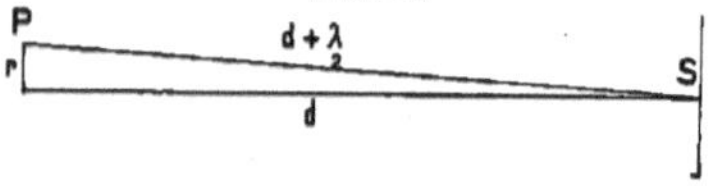

$$r^2 = (d + \frac{\lambda}{2})^2 - d^2;$$
$$= d^2 + d\lambda + \frac{\lambda^2}{4} - d^2;$$
$$= d\lambda \text{ (as } \frac{\lambda^2}{4} \text{ is very small);}$$
$$r = \sqrt{d\lambda};$$
$$\text{or} \quad d = \frac{r^2}{\lambda}.$$

Example.—Let the centre of the plate be 10 inches from the hole, and taking the light near the G line of the spectrum as the most

* Captain Abney on "Pinholes," *Camera Club Journal*, May 1890.

FORMATION OF IMAGES

photographically active, $\lambda = .000017$; the radius of the hole should be :—

$$r = \sqrt{.000017 \times 10}$$
$$= \sqrt{.00017}$$
$$= .013 \text{ inch};$$

or ·026, or $\frac{1}{38}$ inch, is the best diameter for the hole.

By applying this formula to needles of known diameter (Helic needles of Messrs. Milward & Son, Redditch) the following results may be useful :—

Number.	Diameter.	Approx. Diam'r.	Distance for plate
1	.046	$\frac{1}{22}$	31″
2	.042	$\frac{1}{23}$	26
3	.038	$\frac{1}{26}$	21
4	.036	$\frac{1}{28}$	19
5	.032	$\frac{1}{31}$	15
6	.029	$\frac{1}{35}$	12
7	.026	$\frac{1}{38}$	10
8	.023	$\frac{1}{44}$	9
9	.02	$\frac{1}{50}$	6
10	.018	$\frac{1}{55}$	5
11	.016	$\frac{1}{62}$	4
12	.014	$\frac{1}{72}$	3

We shall presently see that diffraction may also have a disturbing effect upon the definition given by photographic lenses, when the opening of the diaphragm bears too small a relation to its focal length.

It may be convenient now to point out that the "pinhole" camera affords a ready means of ascertaining with some accuracy the focal length of positive lenses or lens-systems. A suitable definite distance between the pinhole and screen being accurately set, a negative is taken of some distant object, the image of which is preferably longitu-

TELEPHOTOGRAPHY

dinally and centrally placed, and of considerable contrast in subject. If this image be very carefully measured in length, its measurement forms a gauge for the determination of the focal length of lenses, by comparison with the images formed by them, a matter of simple

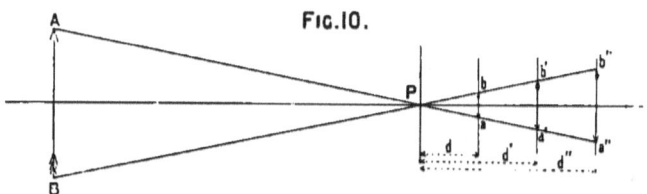

Fig. 10.

proportion—*e.g.*, if the object, A B formed an image, $a'b'$, with a pinhole at the distance d', then two lenses, forming images $a b$, $a'' b''$, would have focal lengths corresponding to the proportion between $a' b'$ and $a b$, $a'' b''$ respectively, or their absolute focal lengths would be d and d''.

PLATE I

Taken with an ordinary 10-in. cabinet lens at a distance of 10 ft.; compare with Plate II. and note the exaggerated size of the hand, flower, and foreground, and the dwarfing of the background. (*By the Author.*)

PLATE II

Taken with an 8¼″ *c. de v.* lens combined with a 4-in. negative lens at a distance of 24 ft. The camera extension was set to render the face in the same scale as Plate I.; the relative planes of the picture have their proper values in respect of drawing.

(*By the Author.*)

CHAPTER III

THE FORMATION OF IMAGES BY POSITIVE LENSES

THE laws of reflection and refraction of light at plane and spherical surfaces are contained in every elementary treatise on "Optics."* The reader will do wisely to master them, although an intimate knowledge of the subject is not necessary in order to understand how images are formed by lenses and combinations of lenses.

For our present purpose, it will only be necessary to understand how pencils of light are affected in their passage through a lens, and to define certain characteristics, or elements as they are termed, of the lens itself.

When a ray of light passes from one transparent homogeneous medium into another (either denser or rarer), it continues its course in a straight line when it meets the surface of the medium perpendicularly (or at "right angles," as it is termed), as A, B, C, D. (Fig. 11.) If the ray encounters the second medium in a slanting (oblique) direction, or at any other angle than a right angle, taking a direction A' B', it alters its direction in passing through the second medium, and is bent, or "refracted," as at B' C', continuing its altered course in a straight line within the second medium. On passing from a second into a third medium, the same process continues as at C' D', and so on. (See Notes.)

The altered direction of the ray will depend upon the character of the medium which it meets in its original course. If the second

* Cole's "Photographic Optics" (Sampson Low & Co.).

TELEPHOTOGRAPHY

medium is denser (or more refractive) than the first, the ray is bent *towards* the perpendicular P P drawn through the point where it meets the second surface as at B′ C′. If, on the other hand, the ray is pursuing its course in a straight line in a denser medium B′ C′ than that which it next encounters, it is bent *away* from the perpendicular P′ P′ drawn through this point as at C′ D′. If one medium be denser (or have a higher refractive index, as it is termed) than another, the more will it bend or refract a ray meeting it at a given obliquity.

In Fig. 11 we have illustrated the passage of a ray of light through

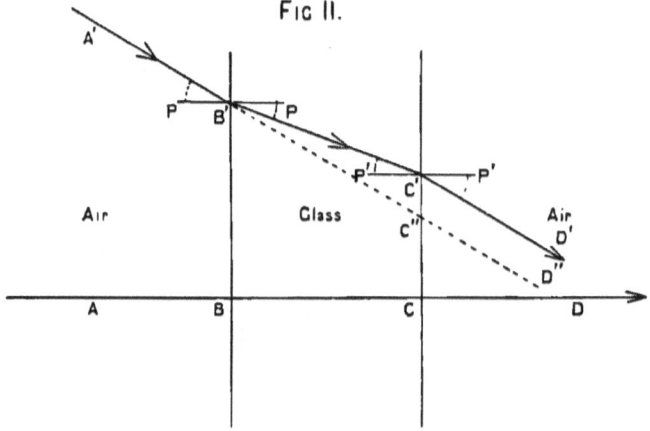

FIG II.

a plane plate of glass with parallel faces, and it will be observed that after passing through the glass it emerges in a direction parallel to its original direction. If A′ B′ had not encountered a medium of different density, it would have pursued the course A′ B′ C″ D″. The effect of the slab of glass with parallel sides has been to divert the ray of light from its course, but not to alter its direction. If the glass is very thick, as in the figure, the ray is considerably diverted; if it be very thin, it is easy to see that its course is very little affected, but in neither case is the *direction* altered. We shall see presently that certain rays pass

FORMATION OF IMAGES

through a lens in precisely the same manner, and shall make use of these rays in our investigations on account of the simplicity of the conditions. It must be noted that the ray originates in one medium (air), and after refraction by the denser medium (glass) emerges again into the same medium (air). This is necessary for the condition; for if the last medium were different from the first, the condition of parallelism would not hold good. In photography the first and final media (air) are almost invariably the same; an exception would occur if we placed the outside surface of the lens under water to photograph fish, &c.

A lens is a portion of a refracting medium bounded by two spherical

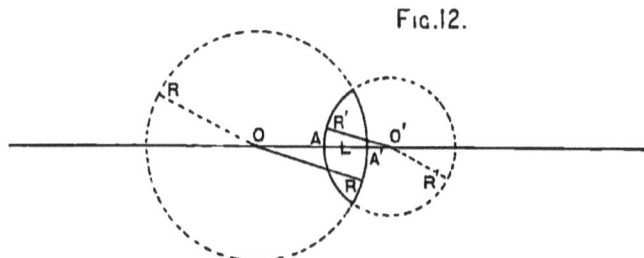

Fig.12.

surfaces of revolution which have a common axis, called the axis of the lens.

With O as a centre and radius O R describe the curve of the spherical surface of revolution R A' R; similarly with O' as centre and radius O' R' describe the curve of the spherical surface of revolution R' A R'. The portion L represents the section of a lens. O and O' are the two *centres of curvature* of the surfaces; O O', the line joining them, *is the axis* of the lens; O R, O' R' the *radii;* A, A', the points where the surfaces cut the axis, are called the *poles*, and the distance between these bounding surfaces, or poles, A A', the *thickness* of the lens. The surfaces of revolution are either spherical or plane—a plane surface being, of course, a particular case of a spherical surface where the radius is infinitely great.

TELEPHOTOGRAPHY

Lenses are classified according to their forms. A lens bounded (1) by two convex surfaces is called a *double-convex* lens; (2) by a convex surface and a plane surface, a *convexo-plane*, or *plano-convex* lens, according to the surface presented to the light; (3) by a convex surface and a concave surface, a *convexo-concave* or *concavo-convex* lens, according to the surface presented to the light. A lens of this last form is also termed a meniscus.

(4) Similarly, a lens bounded by two concave surfaces is called a *double-concave* lens; (5) by a plane surface and a concave surface, a *plano-concave*, or *concavo-plane* lens, according to the surface presented to the light; (6) by a concave surface and a convex surface, a *concavo-*

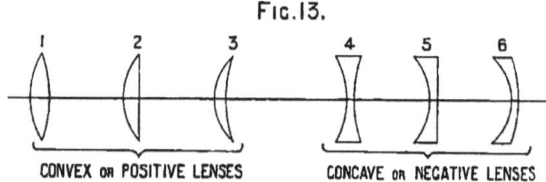

Fig. 13.

CONVEX or POSITIVE LENSES   CONCAVE or NEGATIVE LENSES

convex or *convexo-concave* lens, according to the surface presented to the light. This form of lens is also called a meniscus.

The first three forms of lenses are thicker in the centre than the edge and are called convex or *positive* lenses, and are capable of forming *real* images. The last three forms of lenses are thinner in the centre than the edge, and are called concave or *negative* lenses; they cannot form real images when used alone, but form imaginary or *virtual* images. We may add here, for a special reason that will afterwards be made clear, that we may conceive a "lens" having two plane surfaces; this lens can be imagined to form an image at infinity.

Let us now proceed to determine *the elements of a lens necessary for assigning the position and magnitude of the image of any object.* They are termed the "four cardinal points" of a lens:[*] the two "principal"

[*] See paper on "The Measurement of Lenses," by Prof. S. P. Thompson, *Society of Arts Journal*, Nov. 1891.

FORMATION OF IMAGES

points—these may usually be taken to coincide in one "optical centre" for our present purpose—and the two "focal" points.

I. To determine the position of the "optical centre," and the two "principal points" of a lens.—Draw any two parallel radii O R, O' R' (of the spherical surfaces) and join R R', meeting the axis of the lens in C. The point C is called the *centre*, or optical centre, of the lens.

Any ray of light which, in its passage through the substance of the lens passes through C, will emerge parallel to its original direction, because the lens will act upon such a ray as a plate of glass with parallel faces. In Fig. 14 (3) we have the same figuring as in Fig. 11;

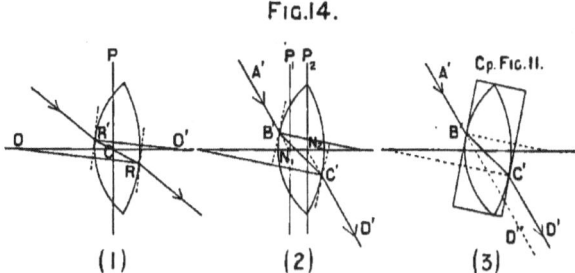

FIG.14.

the direction of A' B' alters in passing through the thickness of the plate, or in passing through the *centre* of the substance of the lens, but the ray emerges parallel to its original direction.

In Fig. 14 (2) if we produce A' B' to meet the axis in N_1, and D' C' to meet the axis in N_2, these two new points, N_1 and N_2, are called the "*principal*," or "*nodal*," points of the lens.

They have the property, which is evident from the figure, that any ray of light proceeding from any direction towards one of these points passes out of the lens as though it had passed through the other. (See Notes to Chapter II.)

We ascertained from Fig. 11 that if the parallel plate is very thin the displacement of the ray passing through it will be very small. The same reasoning will apply to a very thin lens. If we neglect the thick-

TELEPHOTOGRAPHY

ness of the plate the ray A' B' will continue in a straight line, and no displacement will occur. Similarly, in Fig. 14 (2) and (3), the thinner the lens becomes, the less will the ray be displaced in passing through it, and the nearer will N_1 become to N_2, until, if we consider the lens of negligible thickness, N_1 and N_2 will coincide (in the centre c), and any ray proceeding from any direction towards the centre c will pass through it in a straight line.

We see then that in thick lenses there are in reality *two principal points* (see Notes); but for the purposes of our discussion it will be less effort to the reader to consider lenses as very thin, and that these two principal points coincide in one principal point or centre.

Fig.15.

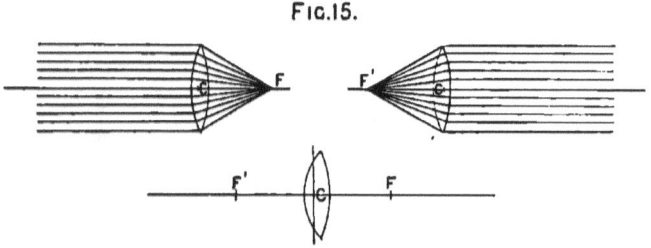

Rays of light proceeding from any direction towards the centre of the lens pass through the centre without deviation.

We define a plane passing through this centre, perpendicular to the axis, as the *principal plane* of the lens.

II. To determine the two focal points of a lens.—If a beam of parallel rays, emanating from a very distant luminous point, meets one surface of a positive lens, in a direction parallel to its axis, the rays will, after passing through the lens, meet in a point F. This point is called the *principal focus*, or principal "focal point," of the lens, and its distance from the centre c of the lens is called the *focal length* of the lens.

Similarly, if we present the second surface of the lens to the incident parallel rays in the opposite direction, they will meet in a point F'.

FORMATION OF IMAGES

This point is the second principal "focal point," and its distance from the centre of the lens C is again a measure of the focal length of the lens.

No real image can be formed at any position nearer to the centre of the lens than F or F'. It is evident that rays from a luminous object placed at either F or F' would, after passing through the lens, emerge parallel; if the object were nearer to C, the rays would *diverge* after passing through the lens.

F and F' are the two *focal points*, and possess the property that *any ray meeting the lens in a direction parallel to the axis of the lens passes through the focal point on the further side of the lens.*

FIG.16.

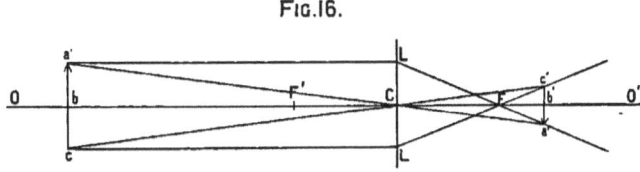

Every lens (or combination of lenses) may be defined as possessing a centre and two focal points.

These elements are sufficient for us to determine the position and magnitude of the image of any object.

We need not now trouble to draw lenses, but indicate them by the principal plane passing through the centre of the lens, and by setting off the positions of the focal points on the axis.

Let O O represent the axis passing through the lens C; and F and F' the focal points (C F = C F').

Let us first determine the *position* of the image of an object *a b c* graphically.

From the object *a* draw a line (or take a ray) parallel to the axis *a* L, meeting the lens at L. This ray, after refraction, must pass through the focal point F in the direction L F *a'*. Now take another ray passing from *a* through the centre of the lens C. This ray passes in a straight

TELEPHOTOGRAPHY

line through c and meets the first ray (which took the direction L F a') in the point a', forming the image of a at a'.

Similarly two rays from c, one meeting the lens parallel to the axis, and the other passing through the centre, will meet in c', forming the image of c at c'.

We have thus determined the position of the image. Next as to its magnitude. If we compare Fig. 3, illustrating the formation of the image by the pinhole camera, with Fig. 16, we observe that the rays passing through the centre of the lens correspond with those passing through the pinhole, and the proportion existing between the sizes of object and image is identical with the relation between the

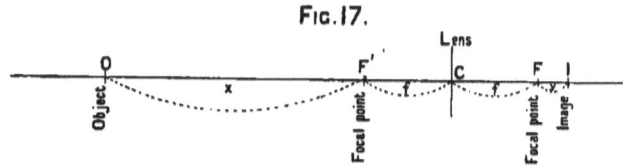

FIG. 17.

distances of object and image from the centre of the lens. The magnitude of the image is to that of the object as c b : c b'.

To express these relations numerically we must introduce an optical law, known as the "*Law of Conjugate Foci,*" which may be deduced from the geometrical construction above.

If we plot c F and c F' as before, calling c F and c F' each $= f$, and calling the distance between F' and the object o, equal to x, and the distance between F' and the image I equal to y; the law tells us that

$$f^2 = xy \quad . \quad . \quad . \quad . \quad (1)$$

which means that the focal length of the lens squared, or multiplied by itself, is always equal to the distance between the object and the front focal point, multiplied by the distance between the image and the back focal point. As this generalisation may at first seem a little confusing to the reader, we will *interpret the law in another fashion.*

Plot c, F', and F as before, measure the distance between the object

PLATE III

Taken with the same $8\frac{1}{4}''$ c. de v. lens combined with a 4-in. negative lens, as Plate II., at a distance of 15 ft. (*By the Author.*)

FORMATION OF IMAGES

O and the front focal point F′, and find what multiple this distance is of the focal length of the lens (in the figure it is five times); the distance between the back focal point F and the image I will be the reciprocal of this multiple of the focal length (in the figure ⅕ of the focal length). O and I are said to be conjugate points, or conjugate to one another. If the object be placed at I the image will be formed at O; so that object and image are interchangeable. The size (linear) of the image will also be one-fifth of the size of the object; in other words, the magnification will be one-fifth.

To take a numerical example:

Supposing the focal length of the lens $f = 10$ inches; then the distance O F′ is 5×10, or 50 inches; the distance of the object from the

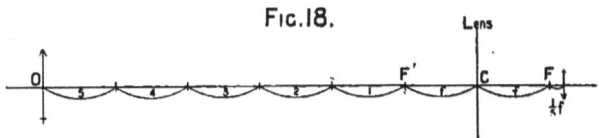

Fig. 18.

centre of the lens is 60 inches; the distance I F is $\tfrac{10}{5}$ or 2 inches, and the distance of the image from the centre of the lens is 12 inches. The size (linear) of the image is to that of the object as their respectives distances from the centre of the lens, or as 12 : 60, that is as 1 : 5; the magnification is one-fifth.

If we substitute the values for the focal length of the lens, the distance from the front focal point to the object, and the distance of the back focal point to the image in the general formula expressing the "law of conjugate foci," we shall find that it holds good.

$$f^2 = xy$$
$$10 \times 10 = 50 \times 2$$
$$100 = 100$$

In general, then, we notice that the further a given object is from the lens, the smaller will be its image, and the more closely will the

TELEPHOTOGRAPHY

image approach to the focal point or plane on the further side of the lens; when the object is very distant, or at "infinity" as it is termed, the image lies in that focal plane.

Conversely, when the object is brought nearer the lens, the plane of the image recedes from the second focal point and increases in size (although the image is smaller than the object), until it arrives at a distance equal to one focal length beyond the front focal point, or, in other words, when it is twice the focal length of the lens away from the centre of the lens. In this interesting position, image and object are of the same size, and are at equal distances from the centre of the lens.

If f be the focal length of the lens, C its centre, F' and F the focal points, and O F' be made $= f$, it follows that F I must $= f$; for O F is now

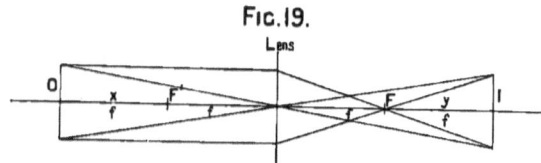

Fig. 19.

the multiple *one* of the focal length, so that F I must be the reciprocal of this multiple, that is to say, *one* also. Or, from the formula:

$$f^2 = xy$$
$$f \times f = f \times f.$$

These particular positions of O and I are called the *symmetric points*, and the planes passing through them the *symmetric planes;* they possess, as we have seen, the property that any object situated in one will be reproduced (but inverted) exactly the same size in image in the other. This is called the position of "unit magnification."

The position of the symmetric points S S' is also remarkable in this respect. If the object is situated beyond one of these on one side of the lens its image is diminished in size (magnification less than unity), and is nearer to the lens than the other: if, on the other hand, the

FORMATION OF IMAGES

object is situated anywhere between one of them and the focal point on that side of the lens, its image is increased in size (magnification greater than unity), and is further from the lens than the other.

Thus, o lying in s is exactly reproduced at 1 in s'; o' to the left of s is diminished in size at 1', and o" to the right of o, between s and F' is increased in size at 1". (Note the interchangeability of object and image in the figure.)

If the object is brought so near to the lens as to coincide with the focal point (or plane through F') we know that, as this point is the position of the focus for parallel rays, the image must be formed at "infinity." Again, if the object be moved nearer still to the lens than this, every ray after passing through the lens will *diverge* and no real

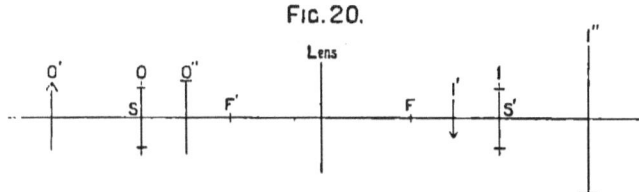

FIG. 20.

image will be formed. (A *virtual* image is formed in this case, but it will not be necessary to enter here into the formation of virtual images by positive lenses.)

It is important for following our subsequent reasoning that the reader should always refer the distance of the object from the lens to a multiple of the focal length of the lens, but bear in mind that the multiple he has to deal with to find the magnification is *one* (*focal length*) *less*. To give another example : suppose the lens to be one of 10 inches focal length as before, and the object he intends to reproduce is 100 inches from the lens. It is distant 10 *times* the focal length of the lens, but only 9 *times* from the front focal point, so that he knows the "conjugate" will be one-ninth of the focal length beyond the back focal point, and also that the "magnification" will be *one-ninth*.

Conversely (and this it should be remarked is the usual manner in

TELEPHOTOGRAPHY

which we shall have to use this law in practice), if we wish to make a certain "magnification" (say one-tenth) with a lens of given focal length (say 10 inches), the distance of the lens from the object must be one focal length *more*, 10 + 1, or 11 times the focal length of the lens.

In general, for a reduction of n times (*i.e.*, a magnification of $\frac{1}{n}$):

Distance of object from lens $= (n+1)$ times the focal length of lens.

Distance of image from lens $= \frac{n+1}{n}$ times the focal length of the lens.

The reader will readily see from the foregoing that as the focal

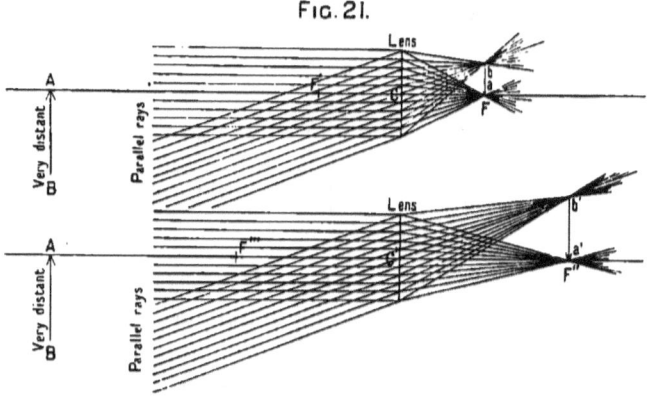

Fig. 21.

points are constant, or fixed, for any given positive lens, this lens can only give larger or smaller images of a given object by placing the lens nearer to it or further from it, in a similar manner to that employed by the pinhole camera with the screen at a fixed distance from the pinhole (Fig. 3). If we wish to produce a larger direct image of an object at a given distance with an ordinary lens the only way will be to employ another lens in which the focal points are situated at a greater distance from the centre of the lens, or one with a greater (or longer) focal length.

FORMATION OF IMAGES

Let Fig. 21 represent two lenses which differ in their focal lengths, that is to say the distances between their centres and focal points differ. In the upper figure the distant object A B will form an image in the focal plane through F, at $a\,b$, of a certain size. In the lower figure the same object will form an image $b'\,a'$ in the focal plane through F'' at $b'\,a'$. It is easy to see that the respective sizes of $a\,b$, $a'\,b'$ are directly dependent on the distances of the focal points from the centres of the lenses, or as C F : C' F''. In other words, the size of an image given by a lens depends upon its focal length; and if one lens has a greater (or longer) focal length than another, the sizes of the images given by them respectively will be directly proportional to their focal lengths.

In general, if one lens is n times the focal length of another, it will give an image n times the size of the other.

We must impress upon the reader that the focal length of a lens is a measurement and definite characteristic of a lens, but he must always bear in mind that this measurement is carried out under definite conditions—viz., that the focal length of a lens is the distance between the centre of the lens and either of the focal points *for parallel incident rays*. This is most important, as confusion might otherwise arise as to the respective sizes of images for objects that are comparatively close.

If we compare the sizes of the image of a very distant object (say a building) given by two lenses, one of which has double the focal length of the other, one image will be found to give an image exactly double the size of the other. If, however, we compare the sizes of the image given by the two lenses of a near object, this relation no longer holds absolutely.

For example, let us compare the sizes of the images given by two lenses of 10 and 20 inches focal length respectively, of an object 100 inches from the centre of the lens :

with the 10-inch lens : $\frac{100}{10} = 10$; $10 - 1 = 9$; magnification $= \frac{1}{9}$;

,, ,, 20- ,, ,, $\frac{100}{20} = 5$; $5 - 1 = 4$; ,, $= \frac{1}{4}$;

so that the sizes of the images are not now as 1 : 2 exactly. The

TELEPHOTOGRAPHY

smaller the multiple of the focal length the object be distant, the more will the relation of the sizes of the images differ from that of their true focal length; if the object be 40 inches distant from either lens,

the 10-inch lens gives a magnification of $\frac{1}{3}$, and

the 20- ,,   ,,   ,,   ,,   ,, 1,

the proportion being as 1 : 3 here.

The usual statement that the right standpoint for viewing a photograph is a distance equal to the focal length of a lens only holds good when all the objects included in the photograph are very distant; it is not then true of all lenses irrespective of their optical construction. For ordinary positive lenses or lens-systems, images of near objects should be viewed as shown above, the correct viewing distance being practically identical with the distance of the conjugate focus of the nearest foreground object in the image from the centre of the lens. But, as we have asked before (p. 13), how are we to know this from the photograph itself?

The only way to evade the necessity of asking the question is to avoid including a foreground that is too near, and to employ a lens whose focal length is considerably longer than the longest side of the trimmed photograph. This practice is both cumbersome and expensive when ordinary positive lenses are employed. The advantages of the Telephotographic lens will be compared presently.

We have, as already stated, endeavoured in this chapter to simplify the study of the formation of images by considering the two principal points of a lens as combined in one, which we have termed the "centre."

If we still keep to this convention, it must be understood nevertheless that this "centre" is not necessarily the mechanical centre of the glass lens in the case of a single lens, nor midway between the two lenses forming a combination of positive lenses, as might be supposed.

If both lenses shown in Fig. 22 are considered thin, the centre c of the equi-convex lens is equidistant from both focal points F F'. In the

FORMATION OF IMAGES

positive meniscus, however, of the same focal length as the equi-convex lens, the centre c is seen to lie outside the lens itself, but its distance from the two focal points F and F' is still the same.

Suppose we place the equi-convex lens in a beam of parallel rays coming from a very distant object, such as the sun, and place a screen in the plane of the image, measuring the distance from the screen to the nearest surface of the lens and noting the size of the image, we shall find that on reversing the lens, the distance between lens and screen is exactly the same and the size of the sun's image is unaltered. In either case the distance between lens and screen is approximately equal to the focal length of the lens. In reality it is slightly less. If the thickness of the lens be considered, one-third of the thickness of the

Fig. 22.

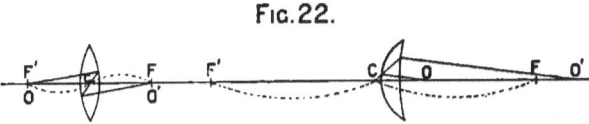

glass lens must be added. Placing now the meniscus lens in the sun's rays, we shall find that if these rays fall upon the convex side of the lens, the distance between the plane of the screen and the nearest surface of the lens is less than in the former case, but the size of the image remains the same. On reversing this lens and presenting the concave surface to the sun's rays, the distance between lens and screen is considerably greater than before, and greater than in the case of the equi-convex lens, but still the size of the image remains the same. The mean of the last two measurements is approximately equal to the focal length of the lens.

We can arrive very approximately at the focal length of either a single lens or a combination of lenses, by two measurements of the distance of the plane of the image (for parallel rays) for some fixed and convenient point in the lens or lens-mount, by presenting first one surface of the lens to the incident rays, and then the other, and taking the mean of the measurements.

TELEPHOTOGRAPHY

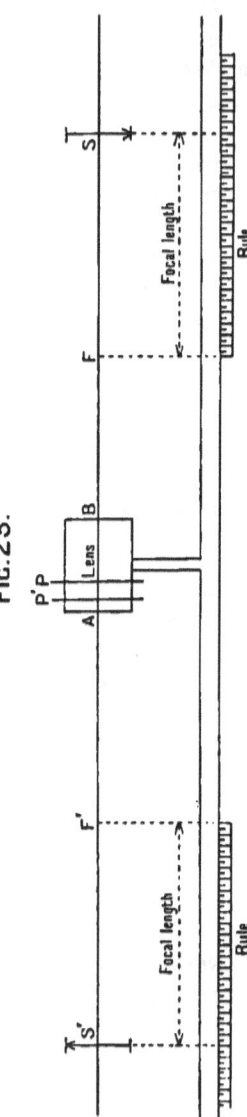

Fig. 23.

We conclude this chapter by giving a method of determining very accurately the true focal length of any positive lens-system without involving calculation, but based upon a principle with which the reader is now familiar.

Mount the lens in a fixed position. Present the surface A of the lens to a beam of parallel rays and thus determine the position of the focal point F. Similarly present the surface B of the lens to a beam of parallel rays, and determine the position of the focal point F'.

Now place a screen at F with a mark or scale of definite length upon it, and a second screen for focusing at F'. Plumbed with the planes through F and F', on the base of the stand, fix two rules, as shown in the figure.

The distance between F and F' is only *approximately* double the focal length of the lens. Remove the screen S, with the definite marking upon it, rather less than half the distance between F' and F to the right of F, and move the focusing screen S' exactly the same distance to the left of F'. An indistinct image of the marking on S will now be seen on S'; proceed to move S and S' small, but exactly equal, distances in opposite directions, until the marking upon S is sharply defined and of equal size on S'. S and S' now occupy the positions for "unit magnifications," and we know that in this position each is removed the true

PLATE IV

Taken with a R. R. lens of 16-in. focal length at a distance of 4 ft., the short distance between lens and sitter has caused the mouth to be exaggerated in size, and the forehead to appear to recede. Compare Plate V. (*By Mr. T. Habgood, Boscombe.*)

PLATE V

Taken with the same 8¼" c. de v. lens combined with a 4-in. negative lens as employed for Plates II. and III., at a distance of 10 ft. The camera extension is chosen to produce the same size of head as in Plate IV. The image could have been made equally sharp, but the back of the positive element was unscrewed to illustrate the soft effect produced by introducing spherical aberration (*see* Fig 58). A picture showing this effect should be viewed at a considerable distance. (*By the Author.*)

FORMATION OF IMAGES

measure of the focal length from the focal points F and F′ respectively. By plumbing at S and S′ we thus read off on the rule the true focal length of the lens.

NOTES.

In Fig. 11 we have illustrated the manner in which a ray of light in passing from air through a plate of glass with parallel sides, emerges finally into air. Let us apply the laws of refraction to the case, in order to show that the incident and emergent rays are parallel to each other.

If we draw a normal P B P′ to the surface of the glass at B at the

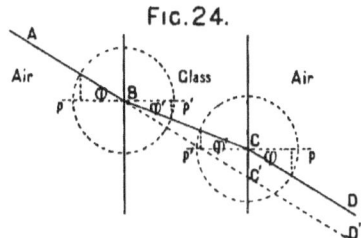

Fig. 24.

point where the incident ray A B meets it, forming an angle ϕ with the normal, we shall find that after entering the second medium the ray will take an altered course B C, and that B C will form a different angle ϕ' with the normal at B. The law of refraction tells us that these two angles, termed the angles of incidence and refraction, are in the same plane, and that the *sines* of these angles bear to one another a constant ratio, depending solely upon the relation of the refractive indices of the two media. Calling the refractive index of air unity, and that of glass the Greek letter μ,

$$\sin \phi = \mu \sin \phi'$$

is a constant for these two media; and in general, in passing from one

TELEPHOTOGRAPHY

medium of refractive index μ to another of refractive index μ', $\mu \sin \phi = \mu' \sin \phi'$.

Glass has a higher refractive index than air (roughly 1.5), thus the angle ϕ' will be less than ϕ.

On meeting the second surface of the glass at C, BC forms the same angle ϕ' with the normal P'CP, and as the ray now passes from a medium whose refractive index is μ into air (refractive index unity),

$$\mu \sin \phi' = \sin \phi,$$

and hence the emergent ray CD forms the same angle Q with the normal as did the incident ray; hence CD is parallel to AB or C'D'.

If the second surface of the plate were not parallel to the first, and the ray formed an angle ϕ'' with the normal to this surface, the angle at emergence, ϕ''', would similarly be found from the equation

$$\mu \sin \phi'' = \sin \phi'''.$$

Let us now apply the laws of refraction to see how rays of light are affected in passing through a lens.

If we trace the passage of a single ray of light, we may always consider it as passing through a plate of glass whose sides are either parallel or inclined to each other. In the latter case, the particular ray passes through the lens as if it were a prism, the ray, on emerging, being always bent or refracted towards its base, or thicker part.

Let us trace the course of two rays parallel to the axis which meet either a positive or negative lens.

The ray AB in either figure meets the lens in the point B; if we draw a line from the centre of curvature of the surface O to meet B, and draw a line P R through this point at right angles to it (termed a tangent), it is evident that the ray AB may now be considered as meeting a plane surface P B R at B, forming an angle ϕ with the normal O B (or perpendicular) to this surface. The ray from B takes an altered course B C making a smaller angle ϕ' with the normal B O, such that $\sin \phi = \mu \sin \phi'$. On the ray arriving at C, join C to the centre of curvature of

FORMATION OF IMAGES

this surface o′, and draw a line through C at right angles to o c, forming the tangent P C S. Before emerging from the lens B C makes an angle ϕ'' with this normal O C N′; after emergence, as it passes from a dense, to a rarer medium, it will be bent or refracted from this normal, forming an angle ϕ''' with it, such that $\mu \sin \phi'' = \sin \phi'''$.

In the case of the positive lens the ray then takes the course C F, meeting the axis in F. F is one of the focal points of the lens, or the locus of its real focus for parallel rays.

In the case of the negative lens, the ray takes the course C D, being bent or refracted from the axis, but proceeding as from a point V on the

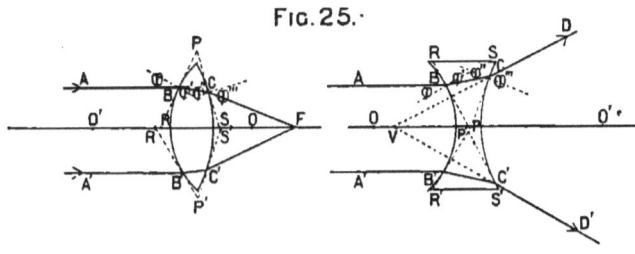

FIG. 25.

axis, in front of the lens. V is one of the focal points of this lens, and is the locus of its *virtual* focus for parallel rays.

The rays thus traced have passed through the lenses, just as though they had passed through the prisms P R S in drawings, being refracted in either case towards the base of the prism.

Similarly, if we trace the course of a parallel ray at a different distance from the axis, such as A′ B′ in either figure, it will be found that its course is the same as through a prism having sides inclined at a different angle, yet finally refracting the ray to the real focal point of the positive lens, and as from the virtual focal point of the negative lens.

From the above it will be easy to see that a ray of light meeting

TELEPHOTOGRAPHY

the lens in any direction will, in general, pass through its curved surfaces as if it passed through a plate of glass with sides more or less inclined to each other; but when the ray in its passage through a lens passes through the centre, the direction of the ray, after refraction, will be parallel to its direction before refraction, or will be affected as if it had passed through a plane plate with parallel sides, as in Fig. 11.

We have drawn attention to the fact (Fig. 14) that there are in reality *two* "principal," or "focal," points in every lens (or combination of lenses), for it is only when a lens is infinitely thin that they can coincide, or become one "optical centre." These two points, and the two focal points, as has already been remarked, are termed the "four

FIG. 26.

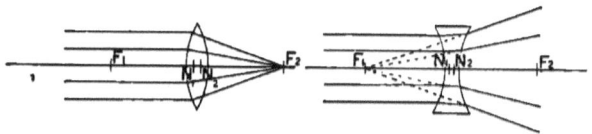

cardinal points" of a lens. In connection with them we define the planes that intersect the four cardinal points in the axis, perpendicular to the axis, as principal and focal planes.

For any given lens the principal points are a certain distance apart, depending on the thickness and form of the lens, and the material of which it is composed, and possess the property that light proceeding from any direction towards one of them passes out from the lens as though it had passed through the other.

If parallel rays are incident upon a lens in either direction, we have seen that F_2 and F_1 are the focal points of the lens, and that their distances from the optical centre of a thin lens are called the focal length of the lens. When the thickness of the lens is taken into consideration, with the *two* principal points, the true measure of the focal length is the distance between N_2 and F_2, or N_1 and F_1. The two focal

FORMATION OF IMAGES

points are always situated at equal distances from their corresponding principal points, or $F_2 N_2 = F_1 N_1$. In other words, any lens or system of lenses has the same absolute focal length whichever surface is presented to incident rays; but it does not follow that the position of the focal points are equidistant from the surface of the lens nearest to them.

Fig. 27 represents a case familiar to many photographers. Here we have a meniscus lens with its principal planes $P_1 P_2$ passing through the principal points $N_1 N_2$ and $F_1 F_2$, the two focal points; $N_1 F_1 = N_2 F_2$.

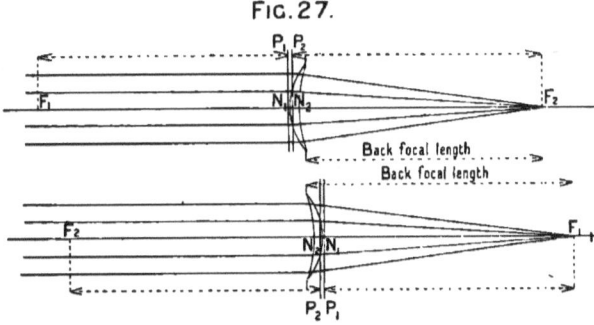

FIG. 27.

It is evident that if the convex side of the meniscus were presented to the light, the distance between the back surface of the lens and the focal point in the same direction would be considerably shorter than if the concave surface of the meniscus were so presented; both images, however, would be of the same size, because the true measure of the focal length is the same in either case, or $N_1 F_1 = N_2 F_2$.

In order to calculate the distances between the principal points for a single lens or combination of lenses, the student is referred to any modern work on geometrical optics. Most positive lenses consist of two combinations. The following formula for calculating the true focal length F, the distance from the back lens to the focal point on the same

TELEPHOTOGRAPHY

side B F, and the resultant width W, between the principal points of the combination, may prove useful:

$$F = \frac{f_1 f_2}{f_1 + f_2 - a};$$

$$BF = \frac{f_2(f_1 - a)}{f_1 + f_2 - a};$$

$$W = w_1 + w_2 - \frac{a^2}{f_1 + f_2 - a},$$

where f_1 and f_2 are the focal lengths of the two combinations, a their distance apart, and w_1 and w_2 the width between their principal points.

To determine the position and magnitude of the image of an object

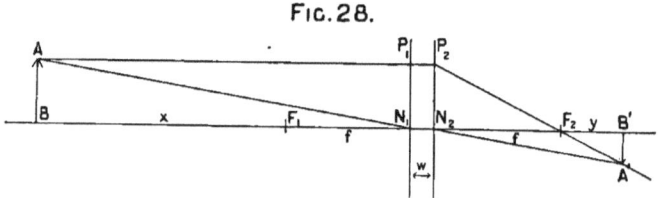

Fig. 28.

formed by a thick lens, or a combination of lenses, we have only to represent the *four* cardinal points and planes of the lens by points and lines as before, setting the principal points and planes at a definite distance apart$=w$ for a single lens, or W for a combination.

A ray from A parallel to the axis passes from P₁ to P₂ in a straight line; from the point where it leaves P₂, it passes through the focal point F₂. Similarly, a ray A N₁, passing towards one of the principal points, N₁, emerges from the other principal point, N₂, in a direction parallel to A N₁, as N₂ A', meeting the former ray in A' and thus determining its position. So that A' B' is the position of the image of the object A B.

FORMATION OF IMAGES

Again $F_1 N_1 = F_2 N_2$, and both are equal to the true focal length of the lens $=f$. Calling $B F_1 = x$ and $B' F_2 = y$; we know that:

$$xy = f^2,$$

and that the size of AB : size of $A' B'$ :: $N_1 B$: $N_2 B'$.

From this it is evident that we need not know the distance between the principal points in order to determine the focal length of a lens, provided we know the positions of the two focal points.

The former method of determining the focal length of a lens illustrated in Fig. 23 will of course apply; but we give another practical method,

FIG. 29.

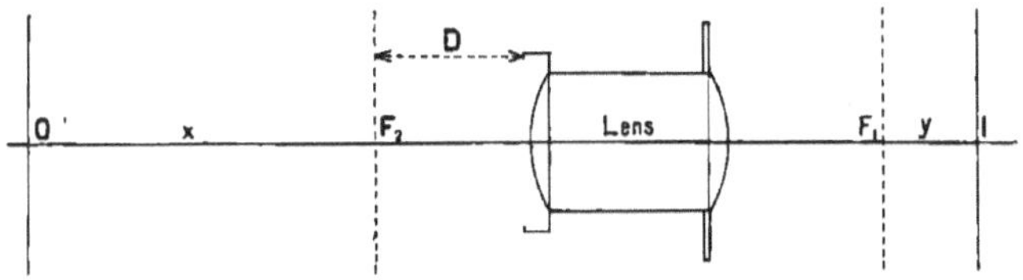

showing that the result can be ascertained without a knowledge of the separation of the principal points.

(1) Determine the position of the focus of the lens for a very distant object upon the screen of the camera and mark the position upon the base-board; this is the plane of the back focal point F_1. (2) Reverse the lens in its flange, and find the position of the focus for the same distant object, measuring the distance of the screen from some fixed point in the lens mount, say the hood. We know that the position of the other focal point F_2 is the distance D from the hood of the lens. (3) Replace the lens in its normal position with the screen at F_1 and then remove it an exact distance y further away—roughly about one-fourth of its distance from the lens for convenience. (4) Find the

TELEPHOTOGRAPHY

distance now necessary for the placing of an object o, so that its image is well defined at the new position of the screen at I. (Operations (3) and (4) may be reversed.)

The distance of the object from the hood of the lens, less D, will be its distance x from the front focal point F_2. Hence $f=\sqrt{xy}$; or, if we multiply x and y together and extract the square root, we find the true focal length of the lens. Supposing $x=2''$, $y=50$, then $f^2=100$, or $f=10$.

CHAPTER IV

THE FORMATION OF IMAGES BY NEGATIVE LENSES

IF we present either surface of a negative or concave lens to a beam of rays parallel to the axis of the lens, we shall be unable to find a real image of the object from which the parallel rays emanated; no real focus will be formed by the lens.

Again, if we bring the object nearer to the lens, until it is in close proximity with it, the lens will still be found incapable of forming a real image.

It is evident, then, that negative lenses, used alone, will be quite useless for photographic purposes.

Our object will be to examine the manner in which rays of light are affected by negative lenses, with a view, subsequently, of using them in conjunction with positive lenses, in such a manner that real images *can* be formed.

Let us first make an experiment with a negative lens to see its effect upon a beam of parallel rays, and acquaint ourselves with the meaning of a "virtual" focus, by determining its approximate focal length in a practical manner.

Take a thin disc of any opaque material, such as cardboard, of the same diameter as the lens under examination; at equal distances from the centre of the disc, and rather nearer the edge than the centre, make two small perforations A, B; measure their distance apart. If we now place this disc in contact with the lens, and present the disc to the sun's rays, we shall find that the pencils passing through A and B

TELEPHOTOGRAPHY

will be divergent after passing through the lens. These divergent pencils can be traced as small discs of light upon a screen held behind the lens; if the screen be made to occupy the position where the distance between these two discs A' and B' is just double the distance between A and B, the distance C F' between the disc and the screen is a rough measure of the focal length of the lens, or is equal to C F. The direction of the pencils A A' and B B' appears to originate at F. F is the "virtual" focus of the lens, and is one of the "focal points" of the lens, F' being the other.

In negative as well as in positive lenses there are two "principal"

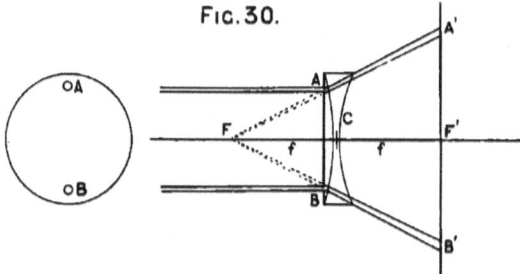

Fig. 30.

points; but here again it will be sufficient for our purpose to consider them as coinciding in one "optical centre."

From the determination of the position of these three elements in a negative lens, we know: (1) That any ray meeting a lens in a direction parallel to the axis emerges from the lens as though it came from the focal point situated on the same side of the lens as the incident ray; (2) that any ray passing through the optical centre of the lens proceeds in a straight line.

We can then, as before, plot a negative lens on paper and indicate it by the principal plane passing through the optical centre of the lens, with its two focal points equally distant on either side.

FORMATION OF IMAGES

Before proceeding, let us again refer to Fig. 30 to see what it teaches us. We have noticed that parallel rays diverge from the lens as though they proceeded from the focal point on the same side. This fact read alone is unimportant to our investigation; we must grasp and remember the converse: that rays (a' a, b' b) converging towards the focal point F of the negative lens on the further side leave the lens in a parallel direction; a focus would be formed by the negative lens at infinity. This is an important, though not practical, conclusion.

Let us now consider the case of an object placed nearer to the lens, but distant some multiple of the focal length (greater than unity).

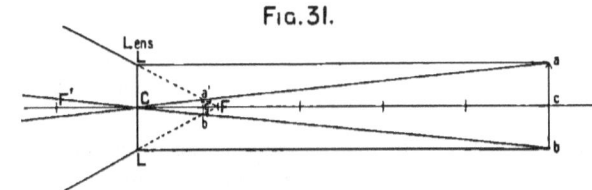

Fig. 31.

Let o o represent the axis passing through c, the optical centre of the lens, and F and F' the focal points (c F = c F'.)

We will now determine the position of the virtual image of the object $a c b$. First consider a ray a L meeting the lens parallel to the axis. This ray after refraction takes the course F a' L, as though the ray emanated from the focal point F. Now take a second ray from a passing through the centre of the lens c; this ray in passing through c cuts the first (virtual) ray at a', and determines the position of its virtual image. In the same manner we can determine the position of the virtual image of b, at b'; $a' b'$ is the virtual image of $a b$, is erect, and lies between the focal point F and the centre of the lens.

To examine the converse: suppose *convergent* rays on the left of the lens are proceeding to form a real image $a' b' c'$, and we interpose the negative lens so that $a' c' b'$ would fall between c and F as shown

43

TELEPHOTOGRAPHY

in the figure, the negative lens will now form an enlarged image of $a'c'b'$ at abc.

In Fig. 31 we have supposed the object acb placed at a distance equal to five times the focal length of the lens, and have seen that the virtual image is formed on the same side as the object.

The relation between the size of object and image is found as follows :

Divide the distance of the object from the lens, by the focal length of the lens, and add one ; this is the *magnification.* (See Notes.)

In the case before us the object is distant from the lens five times its focal length, hence the "magnification" is $5+1$, or 6, or the virtual image is one-sixth the size (linear) of the object. Further, it is evident that the distance of the object from the lens cc is in this same relation to the distance of the image cc'; so that cc' is one-sixth of the distance cc.

Again interpret the converse, for this is the chief interpretation we use :

If *convergent* rays on the left of the lens are proceeding to form a real image at $a'c'b'$, and we interpose a negative lens of known focal length, and find that a real image is formed at acb, we know that the distance of acb from the lens, divided by the focal length of the negative lens, plus one, gives us the magnification of the image $a'c'b'$ occurring at acb.

From Figs. 30 and 31 we see that as the object approaches the lens from infinity the size of the virtual image increases and approaches the lens. If the object be brought up to the plane coinciding with the focal point F, or even nearer to the lens, the same order of things continues, until the object is in contact with the lens, or coincides with the centre of the lens; here image and object both coincide, and we have the position for unit magnification for a negative lens.

To impress this, as the case differs from a positive lens, we will illustrate the case when the object is nearer to the lens than its focal point on the same side.

Taking our two rays whose paths are known :—a L, parallel to the axis, emerges as though it originated at F, and a C passes through the

FORMATION OF IMAGES

centre of the lens; they cut one another at a', forming the virtual image of a. Similarly b' is the virtual image of b, and $a'b'$ that of ab.

Our conclusion from the above observation is: that a real object placed anywhere between the lens and infinity forms a virtual image somewhere between the centre of the lens and its focal point on the same side. And conversely: *Convergent rays incident upon a negative lens, proceeding to a focus situated anywhere between the centre of the lens and its focal point on the other side, will form a real image somewhere between the centre of the lens and infinity on that side.*

If, however, convergent rays meeting a negative lens are proceed-

FIG. 32.

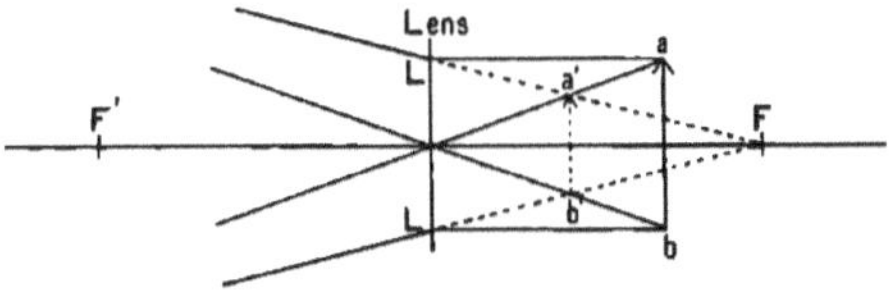

ing to a focus anywhere beyond the focal point on the other side no real image can be formed by the negative lens.

If we examine the convergent pencils (2), Fig. 33, proceeding towards the focal point F, we see that they emerge parallel to the axis, to form an image at infinity. It is easy to see that if they converged to a point beyond F they would no longer emerge parallel, but become *divergent*, and thus could not form a real image.

As a matter of interest let us now illustrate the facts arrived at in a comparative manner.

In (1) the rays converge towards a point anywhere *between* the lens and the focal point F; a real image is formed at R.

In (2) the rays converge towards the focal point F and emerge parallel.

In (3) the rays converge towards a point I, *beyond* the focal point F;

TELEPHOTOGRAPHY

they diverge, as from its conjugate, a virtual image at v, and no real image is formed.

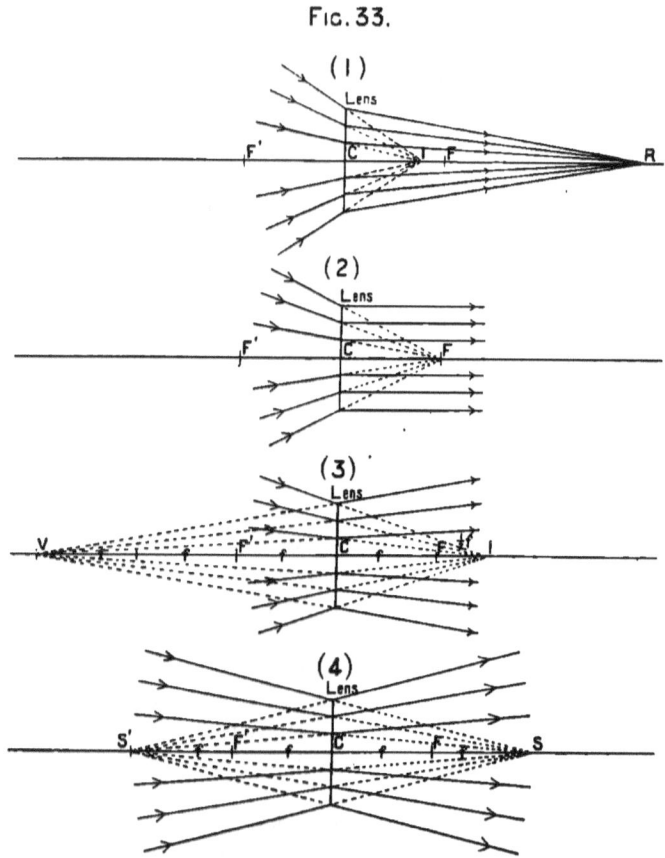

Fig. 33.

In (4), similarly, the rays converge towards the symmetric point s (beyond the focal point), diverging from the lens at the same inclination

FORMATION OF IMAGES

as the incident rays, and as though they originated in the virtual symmetric point S'.

If we reverse (3) we see the conditions for convergent rays proceeding to a point beyond the symmetric point, and in general confirm our conclusion that no real image can be formed by a negative lens when the rays converge to a point beyond its focal point on that side of the lens.

We give in conclusion a practical method of determining the focal length of a negative lens with accuracy:—

1. Focus accurately the image of some well-defined object with any ordinary positive lens, and measure the size of the image.

2. Place the negative lens, the focal length of which we wish to determine, a short distance within the convergent beam from the positive lens. Then focus the image accurately formed by the combined lenses and *measure its size*. Note:

(a) The distance of the screen from any fixed point on the mounting of the negative lens; call this D.

(b) The magnification that has occurred to the image formed by the positive lens alone; call this M.

3. Now move the negative lens a little nearer to the positive lens (N.B.—Keep the positive lens in a *fixed* position), and focus accurately a second time upon the screen, and as before note:

(c) The distance of the screen from the same fixed point on the mounting of the negative lens; call this D'.

(d) The magnification of the image now formed by the positive lens alone; call this M'.

The focal length of negative lens $= \dfrac{D' - D}{M' - M}$.

NOTES]

The four "cardinal points" of a negative lens, as in the case of a positive lens, are the *two* principal points, and the two focal points. Here again the principal points are a certain distance apart, dependent upon the thickness and form of the lens, and possess the property that light proceeding from any direction towards one of them

FIG. 34.

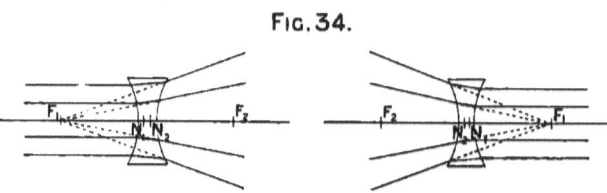

passes out from the lens as though it had passed through the other. The relation $F_1 N_1 = F_2 N_2$ in Fig. 34 always holds good, so that we may trace the course of rays through a thick lens by plotting the two

FIG. 35.

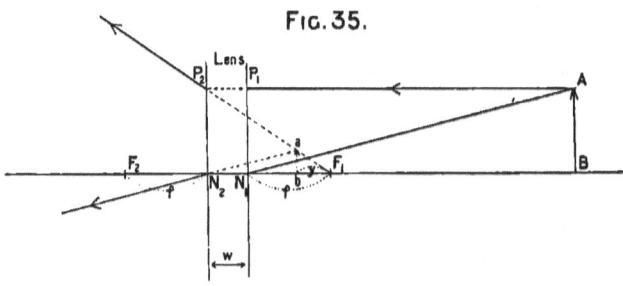

principal planes passing through the two principal points in the axis, and the two focal points on either side.

A ray $A P_1$ passing from A parallel to the axis (Fig. 35) meets the focal plane $P_1 N_1$ in P_1 and emerges from P_2 as though it originated at F_1: similarly the ray $A N_1$ meeting the principal point N_1 emerges from N_2 in a direction parallel to $A N_1$—*i.e.*, in the direction $a N_2$; the point a, where

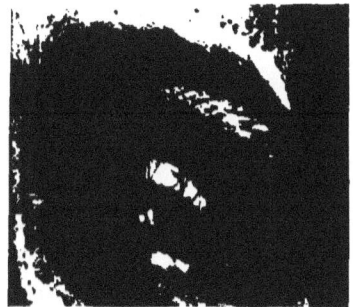

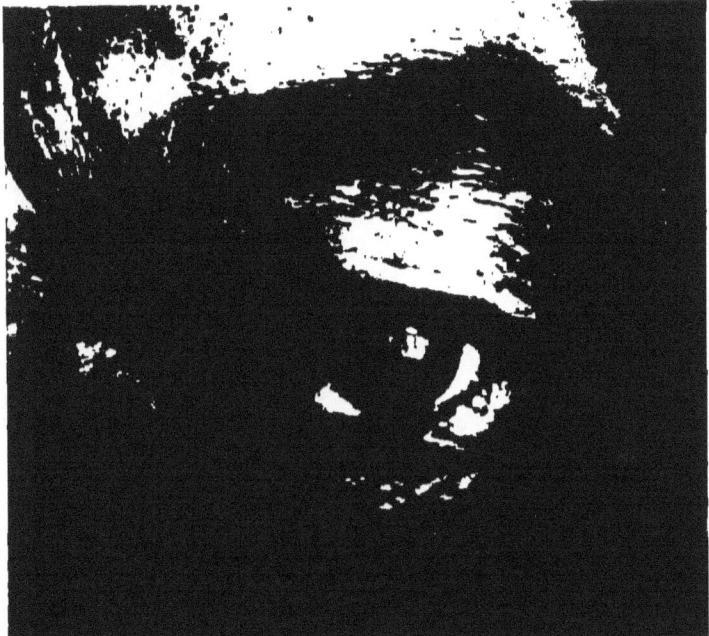

PLATE VI

These three photographs are all untouched, the normal-sized eyes are those of children. Taken with the same Telephoto combination as Plates II., III., and V., at a distance of 4 feet in all three cases. The magnified adult eye was obtained by using a greater extension of camera. Exposure, small eyes, 3 seconds in an ordinary room; large eye 25 seconds in a badly lighted studio. (*By the Author.*)

FORMATION OF IMAGES

N_2 *a* meets P_2 F_1, determines the position of the virtual image *a* of the object A.

In the same manner we can find the image of any other point in the object A B, and thus determine the position of the entire virtual image *a b*.

The relation between the size of the object A B and its image *a b* is found by our interpretation of the law of conjugate foci :

$$xy = f^2.$$

As image and object are here both on the same side of the lens $x = N_1 B + f$, and $y = b F$; so that when $x = n + 1$ times f, $y = \dfrac{1}{n+1} f$.

To take a numerical example : suppose $f = 2$ inches, and the object is 10 inches from it; here $n = 5$ times, so that

$$x = (5 + 1) 2 \text{ and } y = \left(\dfrac{1}{5+1}\right) 2 ;$$
$$f^2 = xy$$
$$= 12 \times \dfrac{1}{3}$$
$$f^2 = 4$$
$$f = 2.$$

Conversely, in examining a real image formed by a negative lens of 2 inches focal length used in conjunction with any positive system, we find this real image is 10 inches distant from the negative lens, and know that it has produced a magnification (very nearly) of $\tfrac{10}{2} + 1$ or 6 times.

In dealing with thick lenses, the measurement should of course be taken from the principal point of emergence, so that a slight error may arise. The position of the principal points in negative lenses differs with their form and thickness.

In the double concave lens A (Fig. 36) both are contained within the glass; in a plano-concave B one is always on the axis where the concave

TELEPHOTOGRAPHY

surface cuts it; and in the negative meniscus, one may fall within the lens or both *may* be outside, as shown in C and D.

In D we note that the ray $a\,b$ proceeding to N_1, and which passes through the centre of the lens (*i.e.*, towards C), emerges from it as $c\,d$ parallel to $a\,b$ and passes through the other principal point N_2.

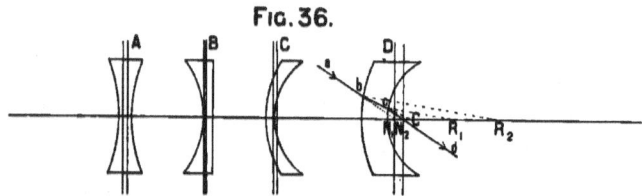

Fig. 36.

In the construction of negative lens-systems used in practice, the principal points do not generally lie outside the lenses, so that only very small or negligible errors come in when determining "magnification" by the above process.

CHAPTER V

THE FORMATION OF ENLARGED IMAGES

PART I. By two Positive lens-systems.
Part II. By a Positive system and a Negative system combined—or the TELEPHOTOGRAPHIC LENS.

PART I.

Every positive lens, as we have seen, is capable of forming a real image. It is evident that this real image may be enlarged by a second

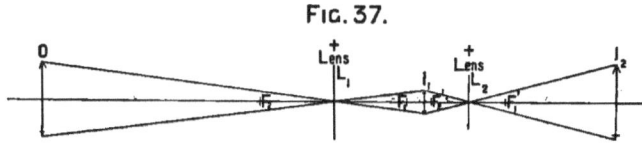

FIG. 37.

positive lens, which in its turn forms a second image of the first one, produced by the first lens.

This process is termed enlargement by "secondary magnification."

The lens L_1 (Fig. 37) forms an inverted image I_1 of an object at O; if we place a second lens L_2 behind this real image I_1 (although formed in air) at any distance greater than the measurement of its focal point F_2 from the lens, a real image of I_2 must again be formed. If the distance between the second lens L_2 and I_1 is greater than the focal length of L_2,

TELEPHOTOGRAPHY

and less than twice its focal length (the position for unit magnification, or the symmetric plane) a magnified and erect image must be formed as at I_2 and can be received upon a screen or photographic plate.

In cases where it is only necessary to magnify a small amount of the image I_1 as occurs in photographing the sun, or portions of it, and where the length and bulk of the instrument is of little moment, this method of magnification has been adopted, although it has recently been abandoned for the "negative" enlarging system. (See Plates VII. and VIII.)

Photographers who know the meaning of "curvature of field" in a lens will see that the curvature of the image I_1 given by the first lens L_1 will be increased by the second positive lens L_2, as the curvature is wrongly disposed for reproduction on a plane surface at I_2.

For general use, however, *bulk* is the great drawback to this system. The first lens necessitates the usual camera extension, equal to or greater than its focal length, and the second lens must increase this by more than four times *its* focal length, before any magnification can begin.

If we were not in possession of the method of enlarging by the negative system, secondary magnification might still be practised in certain cases, because of the advantage gained by the fact that the enlargement given by the second positive lens is of an image formed *in air*. This image has no granular structure, such as takes place in the photographic film on the plate, hence the enlargement would have that "pluck" and definition which are noticeably wanting in enlargements made from the *photographic* primary image.

This leads up to saying here that the whole *raison d'être* for any optical enlarging system is due to the fact that the grain of the photographic image puts a limit, and a very small one too, upon the number of times it can be enlarged with the requisite degree of definition for analysis.

It may be argued that we can see fine definition in the enlargement of a lantern-slide, for example, thrown upon a large screen. So we can as an illusion, when viewing it from a distance, but as we approach the

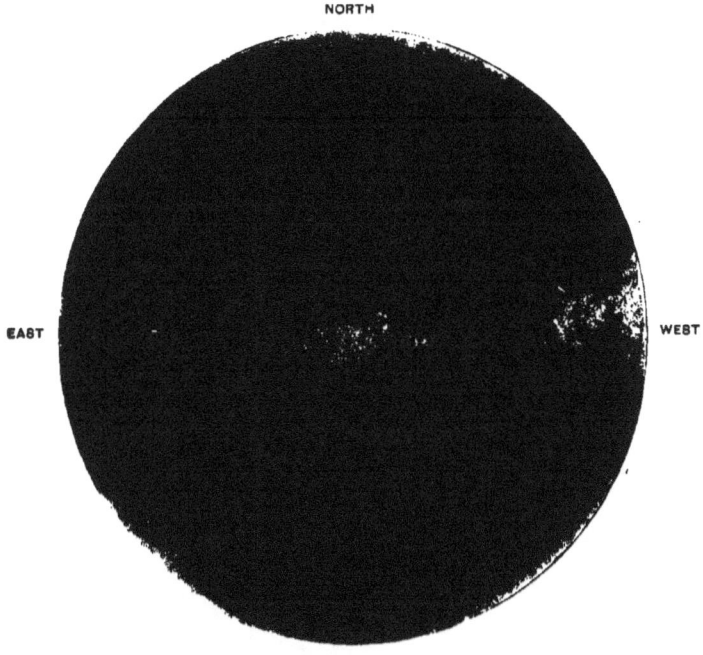

PLATE VII

"Photograph of the great Sun Spot of September 1898. Taken with the Thompson photographic refractor, 26 in. in diameter, aperture employed 15 in., focal length 27 ft. 6 in. Image of the Sun in primary focus 2½ in. in diameter. Enlarged in telescope by a Dallmeyer telephoto lens to 29 in. Taken September 11, 1898."

(*Notes from the Royal Observatory, Greenwich.*)

FORMATION OF ENLARGED IMAGES

screen, all definition is gone! The chief aim and use of optical enlarging systems for taking photographs is to attain as fine definition in the enlarged image as we should in the small image produced by the "positive" portion of the system; in fact, not a *comparative* degree of definition or sharpness, but an *absolute* degree.

PART II.

The Telephotographic lens may be conceived to act in either of two ways—(A) as a *complete positive system* of *variable focal length*, and therefore capable of producing images of *different size* of a given object at any definite distance from it; (B) as consisting of *two separate parts, a positive lens of definite focal length*, whose function is to form a real image of definite size at a definite distance from the object, *combined with a negative lens of definite focal length*, whose function is to *magnify* the image given by the positive lens in *variable degree*.

FIG. 38.

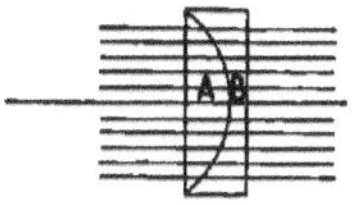

Both conceptions of its action will be dealt with, as they will assist the reader in completely mastering its use in practice. The former will be found the more elegant perhaps in theory, but the author has found that the latter is more easily grasped by the photographer who is fairly intimate with the working of ordinary positive systems.

Let us take two lenses of the same kind of glass, one a planoconvex and the other a concavo-plane, the curved surface being of the same radius in each case, and place them in contact, as in the figure. As the radii are the same, the focal length of each lens is also the same; but as one is positive and the other is negative, the combination acts as a plain disc of glass with parallel sides.

We have seen that we may consider the disc as a lens of infinite focal length, and hence we know that the combination of the two lenses A and B (Fig. 38) forms a compound lens of infinite focal length.

These two lenses, in this condition, form the simplest conception of

TELEPHOTOGRAPHY

the Telephotographic lens that we can imagine, with the components here arranged so as to give the *greatest possible focal length*.

Let us examine this arrangement, so that we may define it in general terms for any other combination of positive and negative lenses used in the Telephotographic construction.

Parallel rays incident upon the lens A (Fig. 39) used alone converge towards the focus F at a definite distance from the lens A; similarly, parallel rays incident upon B diverge as from the virtual focus F', F' being the same distance from the lens B as both have the same focal length.

It is evident that when F_1 is made to coincide with F_2, incident parallel rays will emerge parallel.

In the particular case we have chosen as a preliminary illustration,

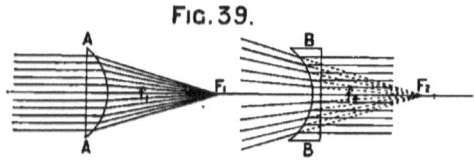

FIG. 39.

on account of its simplicity, as the focal lengths of the two lenses are identical, they must, as is here seen, be placed in contact; but in general *the focal length of the compound lens F is infinite, when positive and negative lenses are separated by a distance equal to the difference of their focal lengths (f_1-f_2)*.

Let us now separate these lenses gradually and observe what happens:

Parallel incident rays upon the positive lens A (Fig. 40) are rendered convergent and proceed to form a focus at F_1; if they are intercepted by the negative lens B separated by an interval a from A they are rendered less convergent, and proceed to form a focus at F, as though they came from A' and not from A. The distance A' F is now the focal length of the compound lens, when the two lenses composing it are separated by the interval d.

FORMATION OF ENLARGED IMAGES

Note.—In general a (see Fig. 40) represents the measurement of the entire separation between positive and negative lenses.

f_1-f_2 is the distance the component lenses must be separated in order that the focal length of the compound lens shall be infinite. (In the present case as $f_1-f_2=0$, the lenses are in contact.)

d represents the measurement of any increase in the separation greater than the difference between the focal lengths of the component lenses f_1-f_2. (In the present case, $d=a$ because $f_1=f_2$.) (See Notes.)

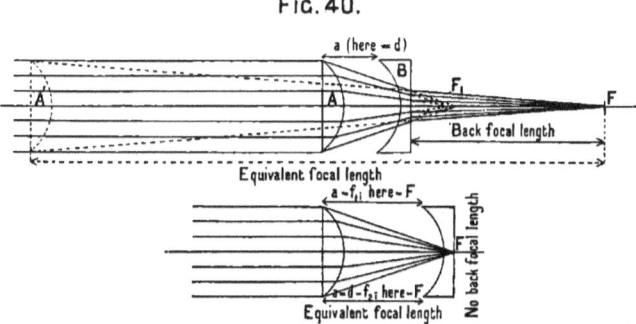

FIG. 40.

If we know f_1, f_2, the focal lengths of the component lenses, a the entire separation, and d the interval as defined, we can find the focal length of the compound lens F, and its back focal length B F (see figure) from the following simple formulæ :

$$\text{F} = \frac{f_1 \times f_2}{d} ; \quad \ldots \ldots \quad (3)$$

$$\text{BF} = \frac{f_2(f_1 - a)}{d} ; \quad \ldots \ldots \quad (4)$$

which tell us that :

The focal length of the compound lens is found by multiplying the

TELEPHOTOGRAPHY

focal lengths of its two components together, and dividing by the interval (d) of separation greater than the difference of their focal lengths.

The back focal length, or distance from the negative lens to the screen, is formed by multiplying the focal length of the negative lens by the difference between the focal length of the positive lens and the entire separation, and dividing by the interval (d) of separation greater than the difference of their focal lengths :

Suppose the lenses A and B, Fig. 40, are both of 6 inches focal length, we see that :

If	d = 0,	F = ∞	B F = 8
	= $\frac{1}{10}''$,	= 360 ;	= 354''
	= 1,	= 36 ;	= 30
	= 2,	= 18 ;	= 12
	= 3,	= 12 ;	= 6
	= 4,	= 9 ;	= 3
	= 6,	= 6 ;	= 0

One of the most valuable practical features of this lens arises from the fact that the back focal length is *shorter* than the true focal length of the compound lens, as illustrated above. This is much more pronounced where the negative lens has a shorter focal length than the positive lens, as will be shown later.

By the above simple combination of two lenses of equal focal length, but of opposite power, we have determined the following general characteristics of any Telephotographic system :

(1) The lens has an infinite focal length when the component lenses are separated by a distance equal to the difference of their focal lengths.

(2) Its focal length decreases from infinity to diminishing finite focal lengths as we increase this separation, until,

(3) When the lenses are separated by a distance equal to the focal length of the positive lens, the focal length of the compound lens is

PLATE VIII

"Photograph of the Eclipse of January 22, 1898. Taken at Sahrdol, India, by the Astronomer Royal. Instrument used, Thompson photographic refractor, 9 in. in diameter, full aperture, focal length 8 ft. 6 in. Image of the Sun in primary focus 1 in. in diameter. Enlarged in telescope by a Dallmeyer telephoto lens to 4 in. Exposure ⅜ second." (*Notes from the Royal Observatory, Greenwich.*)

FORMATION OF ENLARGED IMAGES

equal to this (see Fig. 40) giving its *shortest* focal length. In other words, until the interval d is equal to the focal length of the negative lens, or the negative lens has been separated from the position for emergent parallel rays (as in 1) by an interval equal to its focal length.

(4) The back focal length is always shorter than the true focal length, except in the positions of its longest and shortest focal lengths; but neither of these two positions is of *practical* value. We cannot have a camera of infinite length, nor should we place the plate in contact with the negative lens.

Let us now take a general case, applying the information we have gathered from our simple particular case, in order to examine conditions that are not included in it.

Let the focal length of the negative lens f_2 be shorter than that of the positive lens f_1. If the lenses are placed in contact, it is easy to see that no real image can be formed, for the combination would then be equivalent to a lens of negative focal length. For lenses in contact, the following simple relation holds good:

$$\frac{1}{F} = \frac{1}{f_1} - \frac{1}{f_2}$$

so that if f_2 is less than f_1, F must be a negative quantity. (See Notes.)

The focal length of the compound lens continues to be negative on separating the lenses, Fig. 41 (1), until the separation a is equal to the difference of their focal lengths, or until we arrive at a separation when $d=0$; here the rays emerge parallel or the focal length is infinite (2), as we have seen.

On increasing the separation (3), or making d a definite quantity, the focal length decreases until $d=f_2$, or the whole separation $a=f_1$, when the image is formed in the centre of the negative lens, and the focal length is a minimum, or equal to the focal length of the positive lens f_1. If we increase the separation still further, (4), the negative lens forms a virtual image of the real image formed at f_1.

TELEPHOTOGRAPHY

These results will be readily understood from our examination of the conditions necessary for convergent rays meeting a negative lens to form a real image. (Chapter IV., Fig. 33.)

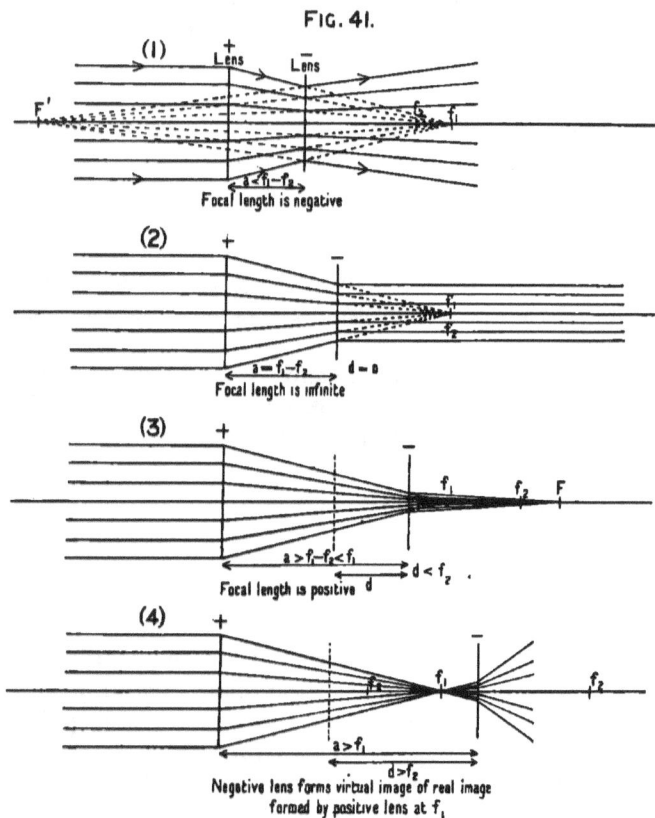

FIG. 41.

In (1) they converge towards a point *beyond* the focal point of the negative lens, and therefore emerge divergent, giving *a virtual image*.

FORMATION OF ENLARGED IMAGES

In (2) they converge towards the focal point of the negative lens and therefore emerge *parallel*; $d=0$.

In (3) they converge towards a point between the negative lens and its focal point, and therefore continue to converge, and form a *real image*; $d=$ a positive quantity greater than 0 and less than f_2.

In (4) the positive lens forms a real image on the axis of the lens in front of the negative lens, which forms a *virtual image* of it.

Diagram (3) of the figure gives us, then, the keynote of the Telephotographic construction: d must be equal to, or greater than 0, and equal to, or less than, the focal length of the negative lens, in order that real images may be formed.

In mounting the lenses of the instrument, we must adjust the separation of the two elements (positive and negative lenses) so that first: the minimum separation makes $d=0$, or parallel incident rays will emerge parallel; and secondly: the maximum separation will allow the negative lens to coincide with the focal point of the positive lens; $d=f_2$. (When mounting the lenses we actually allow a greater separation than this, to provide for the temporary increase in the conjugate focal distance of the positive lens when *near* objects are focused upon.)

In treating the Telephotographic lens as a complete optical system, it is evident that if we make $d=0$ as a starting-point, and know the focal lengths of the component lenses f_1 and f_2, we know the focal length of the compound lens for any value of d greater than 0; for

$$F = \frac{f_1 f_2}{d} \quad \ldots \ldots \ldots \quad (3)$$

that is to say, the focal length of the compound lens is always equal to *the focal length of its two components multiplied together, and divided by the interval d.*

Supposing we mount two lenses of 6 inches positive and 3 inches negative focal length respectively in a tube 3 inches apart (*i.e.*, separated by a distance equal to the difference of their focal lengths), making this their minimum separation ($d=0$), the compound lens has an infinite focal length.

TELEPHOTOGRAPHY

If we now separate them by any greater interval d, say $\tfrac{3}{4}$ of an inch, the focal length

$$F = \frac{3 \times 6}{\tfrac{3}{4}} = 24 \text{ inches.}$$

To mount these two lenses 3 inches apart involves in reality a knowledge of the positions of the principal (or nodal) points of each combination, and would be a difficult matter for an amateur to accomplish. The reader will readily see, however, that if we know the focal lengths of the component lenses, and separate them until parallel incident rays emerge parallel, we have discovered the position in which they are 3 inches apart, without the knowledge of the principal points of either combination.

To take $d = 0$ as a starting-point constitutes the beauty of the method of treating the instrument as a complete system. We can read off the interval d on the mounting of the instrument and instantly calculate the focal length of the compound lens for that particular interval.*

As in the case of ordinary positive lenses, we have now to determine the position of the "cardinal points" of the Telephotographic system, in order to find the position and magnitude of the image of any object.

This is somewhat more complicated, but the reader will not find it difficult to commit to memory the few formulæ necessarily given.

We may consider both component lenses L_1 and L_2 as infinitely thin, but as the principal points of the whole system are sometimes widely separated, we cannot consider them as coinciding in one "optical centre."

We may here state that the focal length of the negative lens is

* The Author devised this method in his early work on the subject, "Telephotographic Systems of Moderate Amplification," 1893, p. 11, engraving the focal lengths of the compound lens for increments of separation, to avoid the necessity of calculation. He has found, however, that the capabilities of the instrument are more readily understood when it is considered as consisting of two separate parts, as described further on, under B.

PLATE IX [E. & H. Spitta, Photos.

"Upper Picture, View of Mattmark Glacier. Photographed with 8.8 Ross Triplet at $\frac{1}{45}$, yellow screen, Edward's iso-medium plate. Exposure three seconds. Top of Glacier about 10 miles distant. The portion included in the telephoto view is that immediately under the two asterisks.

"Lower Picture, Telephoto—The top of the Glacier. Dallmeyer 2B patent portrait lens and high power tele-attachment. Camera extension from back of negative attachment 19 inches. Portrait lens was closed to $\frac{1}{29.6}$. Edward's snap-shot iso-plate with yellow screen. Exposure three seconds. Hour about 9 A.M. Point of View about 10 miles distant.

(*Dr. E. Spitta's description.*)

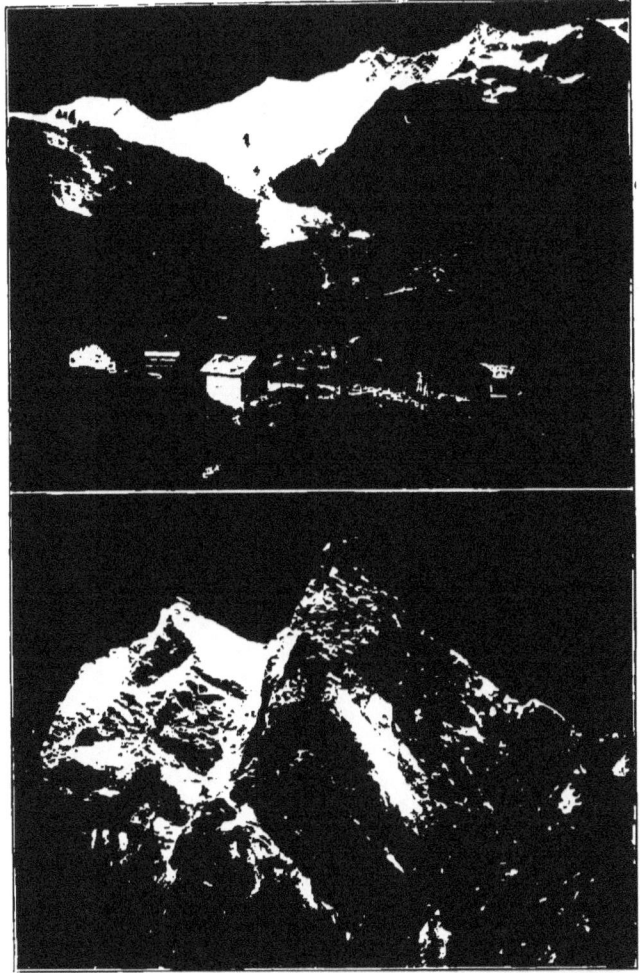

PLATE X [*E. & H. Spitta, Photos.*

"Upper Picture, View of the Saas Grät from Saas Fée. Photographed with Ross 3-inch portable symmetrical $\frac{F}{64}$ yellow screen. Edward's iso-medium plate, exposure three seconds. Hour 8.30 A.M. Mountain about three miles distant. The portion included in the telephoto view is that immediately under the two asterisks.

"Lower Picture, Telephoto—The Döm from other side of Saas Valley. Dallmeyer 2B patent portrait lens and high power tele-attachment. Camera extension from back of negative attachment 20 inches. $\frac{F}{16}$ on portrait lens. Edward's iso-medium plate with yellow screen. Hour 9 A.M. Exposure 12 seconds. Mountain 4½ miles distant." (*Dr. E. Spitta's description.*)

FORMATION OF ENLARGED IMAGES

always shorter than that of the positive lens, hence we may represent f_1 as being a multiple of m (greater than one) of f_2 ;

$$m = \frac{f_1}{f_2} \qquad (5)$$

$$\text{or } f_1 = m f_2. \qquad (6)$$

Let L_1 and L_2 represent the positive and negative lenses respectively, Fig. 42, and d the interval. If we plot f_1, the focal point of the positive lens to the left of the lens L_1, and f_2, the focal point of the negative lens to the right of the lens L_2, it can be shown that [*] :

(1) The front focal point of the whole system F_1 is distant from the front focal point of the positive lens F_1 to $f_1 = m F$;

(2) The back focal point of the whole system F_2 is distant from the back focal point of the negative lens F_2 to $f_2 = \frac{F}{m}$. (See Notes.)

Hence the distance between the front focal point of the whole system and the front lens:

$$F_1 L_1 = m F + f_1 ;$$

and the distance between the back focal point of the whole system and the negative lens:

$$F_2 L_2 = \frac{F}{m} - f_2.$$

In the case of an ordinary positive lens of focal length f, we found that the position of its focal points were each equal to f on either side of the lens respectively ; again, for a given "magnification" of $\frac{1}{n}$: calling the distance of the object from the lens o, and the distance of the image from the lens i,

$$o = nf + f;$$

$$i = \frac{1}{n}f + f.$$

For a similar "magnification" with the Telephotographic lens,

[*] Czapski, "Theorie der Optischen Instrumente nach Abbé."

TELEPHOTOGRAPHY

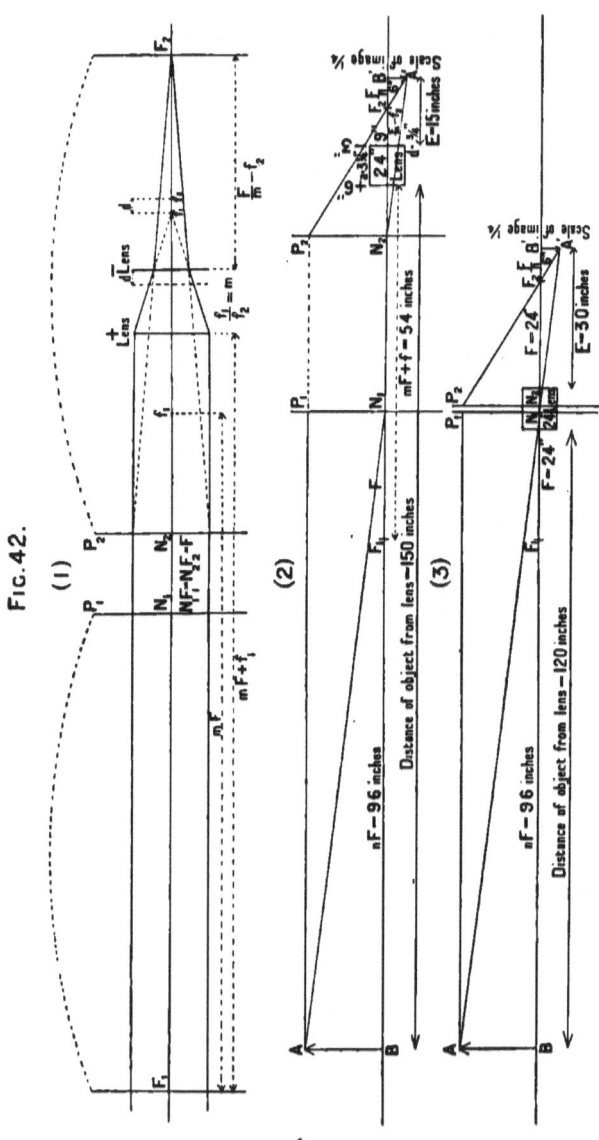

Fig. 42.

FORMATION OF ENLARGED IMAGES

calling o the distance of the object from the positive element and i the distance of the image from the negative element,

$$o = n\,\text{F} + m\,\text{F} + f_1; \quad \dots \dots \quad (7)$$

$$i = \frac{1}{n}\text{F} + \frac{1}{m}\text{F} - f_2 \quad\quad (8)$$

To compare these results, let $f = \text{F}$; hence

$$n\,\text{F} = o - \text{F}; \text{ and}$$

$$\frac{1}{n}\text{F} = i - \text{F}.$$

If we neglect the separation between positive and negative lenses,

$$o = o - \text{F} + m\,\text{F} + f_1;$$

$$i = i - \text{F} + \frac{1}{m}\text{F} - f_2;$$

or, written in another form:—

$$o = o + \left(\text{F}\left(m - 1\right) + f_1\right). \quad \dots \quad (9)$$

$$i = i - \left(\text{F}\left(1 - \frac{1}{m}\right) - f_2\right). \quad \quad (10)$$

From the last two equations* it is evident that using a Telephotographic lens in place of an ordinary positive lens of identical focal length:

(1) The distance of the object must be greater than with an ordinary positive lens for the same "magnification"—an advantage in perspective.

(2) The distance of the negative lens to the plane of the image (or camera extension) is smaller—an advantage in mechanical means.

* D. P. Rudolph, "Gebrauchsanleitung fur Tele-Objective," May 1896.

TELEPHOTOGRAPHY

From the quantities contained in the brackets in equations (9) and (10), it is evident that the greater we make m the more pronounced will the advantages of the Telephotographic lens become with respect to convenience of usage, and in respect also of the better perspective given by a greater distance between lens and object for a given "magnification."

Note.—Equations (7) and (8) must be remembered and applied in conjunction with the general formula $F = \dfrac{f_1 f_2}{d}$ (which determines the focal length of the compound system) to find the necessary distance of object from the front lens of the system for a given magnification, and to ascertain the requisite camera extension.

This method, although elegant in theory, necessitates further calculation to arrive at the *intensity* of the equivalent or compound lens for the time being. This may readily be done by making the diaphragm apertures of definite measurement, in contradistinction to "relative apertures," by dividing the focal length of the lens (for a given interval d) by the aperture used.

When the object is very distant, or as we say at infinity, it is evident that no advantage is derived from the property of the Telephotographic lens shown in equation (9), although the advantage derived from the condition in equation (10) always exists. Referring to equation (3) when n is small, or of the same order as m, then we find the increased distance between object and lens a great advantage, as will be seen when dealing with the subject of portraiture; but when n is great, then m becomes insignificant in comparison, and no advantage is practically attained.

Having ascertained the position of the two focal points $F_1 F_2$, let us proceed to find the position of the principal points $P_1 P_2$, Fig. 42.

It is evident that their distance apart,

$$P_1 \text{ to } P_2 = F_1 \text{ to } F_2 - 2F.$$

FORMATION OF ENLARGED IMAGES

The whole distance

$$F_1 \text{ to } F_2 = mF + 2f_1 + d - 2f_2 + \frac{1}{m}F;$$

$$= \frac{f_1^2}{d} + 2f_1 + d - 2f_2 + \frac{f_2^2}{d};$$

$$= \frac{1}{d}(f_1 - f_2 + d)^2 - (\frac{2f_1f_2}{d});$$

but $2F = \frac{2f_1f_2}{d}$,

Hence $P_1 \text{ to } P_2 = \frac{1}{d}(f_1 - f_2 + d)^2$ \hfill (11).

P_1 and P_2 are equidistant from F_1 and F_2, as in any positive system, and thus their positions and width are determined.

Equation (11) shows us that when the interval d is small, the principal points are *widely* separated, so that with the Telephotographic lens we cannot consider them as coinciding.

Let us now see how to determine the position and size of the image of an object.

Plot the two principal points of the lens $P_1 P_2$ (which here lie outside the lens) and their corresponding focal points $F_1 F_2$ (see Fig. 42 for convenience of following the disposition of the cardinal points). Let A B be an object n times the focal length of the combination distant from the focal point F_1, thus $B F_1 = nF$. The ray $A P_1$, parallel to the axis, passes from P_2 through the focal point F_2 in the direction $P_2 F_2 A'$; the ray $A N_1$, proceeding to the front principal point N_1, proceeds from the second principal point N_2 in a direction parallel to $A N_1$, or in the direction $N_2 A'$, meeting the first ray in A'. A' is then the image of A; and $A' B'$ the image of A B. As $B F_1 = nF$, we know from our previous study of a positive lens, that $B' F_2 = \frac{1}{n}F$. Again the size of
A B : A' B' :: n : 1.

Let us now give a numerical example comparing a Telephotographic lens with an ordinary positive system of the same focal length. (See (2) and (3) Fig. 42.)

TELEPHOTOGRAPHY

Let the focal length of the ordinary positive lens be 24 inches.

Let the focal lengths of positive and negative lenses of the Telephotographic system be 6 and 3 inches respectively; $\frac{6}{3} = 2$, or m here $= 2$. To make this equivalent to the 24-inch lens, the interval d must be $\frac{3}{4}$ of an inch: $\frac{6 \times 3}{\frac{3}{4}} = 24$.

Suppose we want to produce an image $\frac{1}{4}$ of the size of the object A B.

In both cases $B F_1 = 4 \times 24 = 96$, and similarly $F_2 B' = \frac{24}{4} = 6$ inches, and therefore $A B : A' B' : : 4 : 1$ in each case.

In the case of the Telephotographic lens the distance of the object from the lens must be $\overset{(n)}{4 \times 24} + \overset{(m)}{2 \times 24} + \overset{(f_1)}{6} = 150$ inches; but with the ordinary lens only

$$\overset{(n)}{(4 + 1)\, 24} = 120 \text{ inches.}$$

Again for camera extension with the Telephotographic lens

$$\underset{(n)}{\frac{24}{4}} + \underset{(m)}{\frac{24}{2}} - \overset{(f_2)}{3} = 15'',$$

but with the ordinary lens

$$24 + \underset{(n)}{\frac{24}{4}} = 30''.$$

The convenience of the shorter camera extension is evident and very marked.

The greater distance between lens and object is of great importance from the point of view of perspective, for the reader must bear in mind that the position and magnitude of the image of an object derived from the four cardinal points, although accurate, does not trace the course of

FORMATION OF ENLARGED IMAGES

those rays which in reality produce the image. In the example above it is evident that the ordinary positive lens, being nearer to the object, includes it under a greater angle than does the Telephotographic lens at a greater distance. When the distance of the object is very great the advantage of course ceases. It may be stated very approximately that : *The distance of the object from the lens determines the perspective.* We shall refer to this more fully later.

NOTES ON A.

If two lenses, f_1, f_2, of positive and negative focal length respectively, are separated by a given distance a between their principal or nodal points, the focal length of the combination

$$F = \frac{f_1 \times f_2}{f_2 + a - f_1}.$$

If we make $d = f_1 - f_2 - a = 0$, $a = f_1 - f_2$; or when the lenses are separated by a distance equal to the difference of their focal lengths $d = 0$ and the focal length of the combination is infinite.

If $d = 0$, or is any interval greater than the difference of the focal lengths, and equal to or less than f_1, the formula takes the form

$$F = \frac{f_1 \times f_2}{d}.$$

When $a = f_1$, $d = f_1 - f_2 - f_1 = -f_2$, or the interval d is equal to the focal length of the negative lens ; the condition when the combination has its shortest focal length ; $F = f_1$.

If f_1 is some multiple of f_2, we may say

$$f_1 = m f_2,$$

where m is the multiple.

Substituting in : $F = \frac{f_1 f_2}{d}$; it is evident that

$$F = \frac{1}{m} \left(\frac{f_1^2}{d}\right);$$

TELEPHOTOGRAPHY

and also
$$F = m \left(\frac{f_2^2}{d}\right).$$

The distance of the front focal point of the combination from the front focal point of the positive lens $= \left(\frac{f_1^2}{d}\right)$ and $\therefore = m\,F$.

Similarly the distance of the back focal point of the combination from the back focal point of the negative lens $= \left(\frac{f_2^2}{d}\right)$ and $\therefore = \frac{1}{m}\,F$, and hence:

Distance of front focal point from positive lens $= m\,F + f_1$.

„ „ back „ „ „ negative lens $= \frac{1}{m}\,F - f_2$.

METHOD B.

Let us now consider the Telephotographic lens as consisting of two separate parts.

The positive element may be considered as forming an image of definite size, of a given object, dependent upon its focal length, and the negative element may be considered as enlarging the image which the positive lens would have formed by itself.

In general, if one positive lens is n times the focal length of another we have seen that the image produced by the one will be n times as large as that produced by the other. From this it is evident that if we increase the size of any given image n times, this enlarged image will be identical with that produced by a lens whose focal length is n times the focal length of the lens that produced the original image.

In Chapter III. we have ascertained how images of either near or distant objects are formed by a positive lens; and in Chapter IV. we found the law for finding the "magnification" of the image of an object by a negative lens: "Divide the distance of the object from the lens, by the focal length of the lens and add one." The image formed by a negative lens alone is virtual, and as the above law shows us that a

PLATE XI

Photograph by 10-in. stigmatic) of **Boscombe Gardens** in foreground and the **Isle of Wight** in the distance. The "Needles" in the centre are not visible, as the scale they are rendered in is too small; the cliff can just be distinguished. Taken 3 P.M., April 19, 1899.

(*By the Author.*)

PLATE XII

Photograph from the same standpoint as Plate XI, and at the same time by telephoto lens, composed of 8½" c. de v. as positive element, and 2-in. negative. Camera extension from negative element 18 in. Exposure seven seconds; yellow screen and isochromatic plate.

(By the Author.)

FORMATION OF ENLARGED IMAGES

diminished virtual image is formed by a negative lens of a real object we must invert the conditions and form a real image of a virtual object! (Image and object are always interchangeable)

This is precisely what we do accomplish in the Telephotographic construction.

The positive lens L_1, of focal length f_1, would form a real image at $a\,b$, but the rays are intercepted by the negative lens L_2 of focal length f_2. We may then consider $a\,b$ a new "virtual object," a real image of which A' B' is formed by the negative lens L_2.

In this method of treating the subject we shall always refer the size of the final image A' B' to the size of the image $a\,b$ which will be formed by the positive lens alone.

We repeat that in speaking of relative sizes, we refer to *linear* "magnification," unless otherwise stated.

In the Telephotographic lens we may place the screen at any distance we choose from the negative lens, and in order to find how many times we have magnified the image formed by the positive lens alone, we have seen that we must divide this distance by the focal length of the negative lens and add one.

Calling M the magnification, and E the camera extension (distance between negative lens and screen), and f_2 the focal length of the negative lens:

$$\mathrm{M} = \frac{\mathrm{E}}{f_2} + 1 \qquad (12)$$

Rule.—To find the magnification: divide the camera extension by the focal length of the negative lens and add one.

Conversely,

Rule.—To find the camera extension necessary for a certain magnification: multiply the focal length of the negative lens by the magnification less one.

$$\mathrm{E} = f_2(\mathrm{M} - 1) \qquad \cdots \qquad (13)$$

These two rules apply for either near or distant objects. For a near object the focal length of the positive lens may be said to be

TELEPHOTOGRAPHY

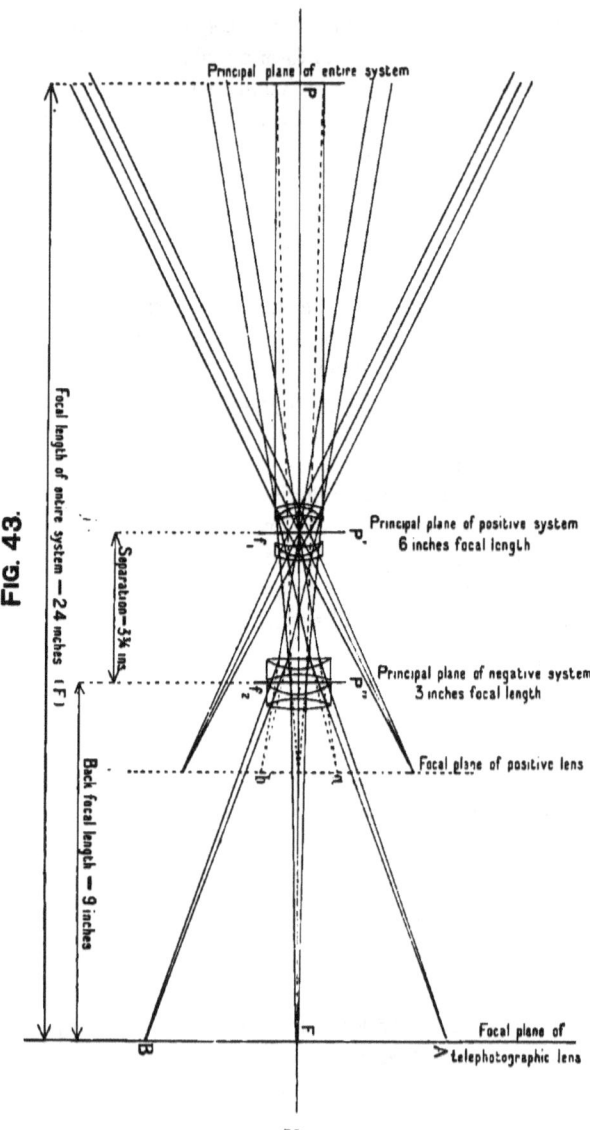

FIG. 43.

FORMATION OF ENLARGED IMAGES

temporarily increased and the image of a near object produced by it is larger than when the object is distant, obeying the law of conjugate foci. The conditions for the magnification of this image are not, however, interfered with.

If the object is very distant (the position of the focus of the positive lens being that of its focal point) we can at once derive the focal length of the Telephotographic combination for any given magnification. It is simply

$$= M f_1 ; \quad F = M f_1 \qquad \ldots \ldots \ldots (14)$$

Rule.—To find the focal length of Telephotographic lens for any chosen extension of camera: multiply the focal length of the positive lens by the magnification.

This may also be found in a perhaps simpler manner from the following consideration. As the positive lens f_1 is some multiple of the negative lens f_2, if we call this m, then $m = \frac{f_1}{f_2}$, and the focal length of the combination for a distant object,

$$F = m E + f_1 \qquad \ldots \ldots \ldots (15)$$

Rule.—The focal length of a Telephotographic lens is equal to m times the camera extension plus the focal length of the positive lens.

Let us now apply these rules to the example illustrated in Fig. 43. Here $f_1 = 6$ inches; $f_2 = 3$ inches; $E = 9$ inches, and $\frac{f_1}{f_2} = m = 2$. Hence,

$$M = \frac{9}{3} + 1 = 4 ;$$

and

$$E = 3 (4 - 1) = 9,$$

or the image is four times as large as that given by the positive lens alone, and it is evident that:

$$F = 4 \times 6 = 24 \text{ (from (14))} ;$$

or again,

$$F = 2 \times 9 + 6 = 24 \text{ (from (15))}.$$

TELEPHOTOGRAPHY

We thus see that with a camera extension of only 9 inches from the negative lens, we can obtain an image of the same size as that given by an ordinary positive lens of 24 inches focal length, requiring this length of camera!

In other words, the true focal length is the distance between P and F, although the distance between P″ and F is all that is necessary to obtain it. As in ordinary positive systems, P is one of the principal points and F one of the focal points of the Telephotographic system.

By treating the lens in this manner we do not necessarily know the separation of the principal points or planes of the two components, and for distant objects the knowledge of the distance between the principal points and planes of the whole system is immaterial. For extreme accuracy in the results, we ought to know the position of the principal point or plane of the negative lens, as it should be from this point that we set the camera extensions. In practice the mechanical centre of the thickness of the negative lens is *very nearly* the position of its principal point, and negligible errors only are likely to creep in.

It is perhaps needless to say that the separation of the two elements determines the plane of the final image, and we have to adjust the separation of the two lenses mechanically until the image is well defined upon the screen in the position we have assigned to it, in order to bring about the required magnification.

Again, by considering the final image as a magnification of the image produced by the positive lens alone, we see, at once, from first principles, how to determine the comparative illumination of the two images, which decides the relative photographic exposures.

Suppose we know the correct exposure to give with the positive lens alone, and we magnify this image by means of the negative lens a certain number of times (linear), the *surface* covered is the square of the linear dimension; hence if t is the correct time of exposure to give for the positive lens alone, the correct exposure for a given magnification (linear) with the Telephotographic illumination is:

$$\text{T, the exposure} = \text{M}^2 t \quad \quad \quad \quad (16)$$

FORMATION OF ENLARGED IMAGES

Rule.—To find the correct exposure with the Telephotographic lens: Multiply the time of exposure necessary for the positive lens alone by the square of the magnification.

This result is identical with the following:

Rule.—To find the *intensity* of the Telephotographic lens: Divide the intensity of the positive lens by the magnification.

We shall refer to intensity in the next chapter, but may point out here that to find the relative exposures of two lenses we have to square the denominators of the fractions expressing them. So that if the intensity of the positive lens alone $I = \frac{a}{f}$, where a is the aperture, and f_1 the focal length, the intensity of the Telephotographic combination for a magnification of M times $= \frac{a}{Mf}$, or the relative exposures are as: $f_1^2 : (Mf_1^2)$ or $f_1^2 : F^2$. And this is the same thing as saying that with a given aperture, or definite size of diaphragm of the lens, the relative exposure for the positive lens alone and the Telephotographic combination are as the squares of their focal lengths.

If we examine Fig. 43, we shall see that only a comparatively small part of the entire image formed by the positive lens is taken up and magnified by the negative lens; the "drawing" of this small area is identical with that of the final image, but of different scale. It is evident that the scale is dependent upon the magnification employed. The magnification is due to a multiple of the focal length of the positive lens, produced at a distance which is only a (nearly equal) multiple of the (shorter) focal length of the negative lens; hence it is evident that the correct viewing distance of the image cannot be from the distance it is removed from the lens, but a considerably greater distance! We shall examine the correct position in the next chapter.

Let us now investigate, more particularly, the employment of the Telephotographic lens in forming images of near objects, an application of great practical interest and importance to photographers.

We have seen that when we require to produce a certain "magni-

TELEPHOTOGRAPHY

fication" with an ordinary positive lens, the following relations exist:

$$\text{Distance of object} = (n + 1)f_1;$$
$$\text{,, \quad ,, image} = \frac{n + 1}{n} f_1.$$

We have now only to decide how many times we wish to increase the size of *this* image to obtain the desired magnification, for we know that the size of the image formed by the positive element is $\frac{1}{n}$, the actual size of the object. Let M represent the magnification given by negative lens, as before, and $\frac{1}{N}$ the entire "magnification" we wish to bring about between the object and final image *by the whole system*.

$$\frac{1}{N} = \frac{1}{n} \times M \tag{17}$$

For equal size $N = 1$; $\therefore M = n$

half ,, $\frac{1}{N} = \frac{1}{2}$; $\therefore M = \frac{n}{2}$;

For quarter size $\frac{1}{N} = \frac{1}{4}$; $\therefore M = \frac{n}{4}$; and so on.

EXAMPLE.—Suppose the focal length of the positive element $f_1 = 10$ inches, that of the negative lens $f_2 = 4$ inches, and the object is 80 inches distant, and we wish to reproduce it in half natural size.

The lens f_1 produces an image $\frac{1}{7}$ the size of the object; we require the final image to be but $\frac{1}{2}$ or $\frac{1}{N} = \frac{1}{2}$; hence we must magnify the image given by f_1, $3\frac{1}{2}$ times, or make $M = 3\frac{1}{2}$.

$$\frac{1}{2} = \frac{1}{7} \times \frac{7}{2} = \frac{1}{2}.$$

applying the formula:

$$E = f_2 (M - 1)$$
$$= 4 (3\frac{1}{2} - 1) = 10,$$

FORMATION OF ENLARGED IMAGES

or we require a camera extension of 10 inches to arrive at the final magnification wanted.

Let us now determine the true focal length F of the Telephotographic system that has produced the result.

It is obvious that one of the focal points F_1, or the position of its focus for parallel rays, must be somewhere between the negative lens and the screen.

If we call its distance from the negative lens $F L_2 = x$, F S must be $\frac{F}{N}$; calling E = the camera extension as before, we have the two following relations:

$$x = E - \frac{F}{N};$$

$$F = mx + f_1;$$

hence

$$F = m\left(E - \frac{F}{N}\right) + f_1;$$

or

$$F\left(\frac{m}{N} + 1\right) = mE + f_1;$$

$$F = \frac{mE + f_1}{\left(\frac{m}{N} + 1\right)}. \qquad \ldots (18)$$

In the example given $m = \frac{10}{4} = 2\frac{1}{2}$, $f_1 = 10$, E = 10, and $\frac{1}{N} = \frac{1}{2}$.

$$F\left(\frac{2\frac{1}{2}}{2} + 1\right) = 2\frac{1}{2} \times 10 + 10;$$

$$1\frac{5}{8} F = 35;$$

$$F = 15\frac{5}{9}.$$

The knowledge of the true focal length is necessary for determining the intensity of the lens, and therefore equation (18) must be remembered and applied. It will be noted that the result obtained is identical with that given by (8).

Knowing that $F = 15\frac{5}{9}''$, the distance between object and image to bring about the magnification of $\frac{1}{2}$ must be (or would be thought to be)

TELEPHOTOGRAPHY

$4\frac{1}{2}$ F or 70 inches; but (neglecting the separation of positive and negative lenses) the whole distance = 80" (distance of object from positive lens) + 10" (camera extension from negative lens = E) or 90 inches, hence the separation of the principal points of the whole system is 90 − 70, or as much as 20 inches!

If we required an ordinary positive lens-system to give the same "magnification" ($\frac{1}{2}$), at the same distance from the object, its focal length would have to be $\frac{80}{3}$, or 27 inches nearly, and the camera extension 40 inches.

The photographer will see at once the advantage gained by using the Telephotographic lens, giving the same size of image at the same distance as the ordinary lens of longer focal length: (1) in respect of rapidity, if equal effective apertures are employed, or (2) in respect of greater "depth of focus," if both lenses are used at the same intensity, the ratio of the effective aperture to focal length being the same in each case.

Equation (18) shows us that when N is very great, the above advantages of the Telephotographic lens no longer hold good, for F $(\frac{m}{N} + 1)$ then becomes F $(\frac{m}{\infty} + 1)$, or is equal to F, and the equation becomes as before: F = m E + f_1. (See Equation (15).)

Let us now proceed to find the distance that must intervene between the object and the Telephotographic lens for a given "magnification" of $\frac{1}{N}$, when its focal length is known.

PLATE XIII   Taken from the end of **Boscombe Pier**, which is 600 yards distant from the small house on the cliff.   The stratification of the cliff immediately below the house is hardly discernible when photographed with an ordinary lens.   A stigmatic 1/10-in. focal length was

PLATE XIV

Taken from the same standpoint as Plate XIII by a telepho lens composed of an 8¼-in. *c. de z.* lens, combined with a 2-in. negative lens. The stratification is here clearly visible. Camera extension 20 in. (*By the Author*).

FORMATION OF ENLARGED IMAGES

In the first place we must determine the distance of the focal point from the negative lens in order that it may correspond to the focal length F chosen, and to this add $\frac{F}{N}$ for the given magnification, thus determining the entire camera extension E that will be required.

Now we must determine the magnification given to the image formed by the positive lens f_1, to bring about the final relation between object and image, as from the size of the image formed by f_1 we at once find the necessary distance for the object to be placed:

as before
$$M = \frac{E}{f_2} + 1,$$
and
$$\frac{1}{N} = \frac{M}{n};$$
or
$$n = MN \quad \quad \quad \ldots \quad (19)$$
and the distance of the object
$$= f_1 (n + 1) \quad \quad cp \ (2)$$

Example: Take the case of the Telephotographic lens illustrated in Fig. 43 where the focal length is 24 inches. To find the necessary distance between object and lens for a magnification of $\frac{1}{5}$.
Here
$$E = 9 + \frac{24}{5} = 13\tfrac{4}{5};$$
$$M = \frac{13\tfrac{4}{5}}{3} + 1 = 5\tfrac{3}{5};$$
$$\frac{1}{N} = \frac{1}{5} \text{ or } N = 5,$$
$$n = \frac{28}{5} \times 5 = 28,$$

or the distance of object = 6 (28+1) = 174 inches.

The principal points are here separated by a distance of 174 − (6 × 24 + 15) = 15 inches.

TELEPHOTOGRAPHY

For an ordinary positive system of the same focal length

Distance of object $= f_1 (N + 1) = 144$ inches.

" " image $= f_1 \left(\dfrac{N + 1}{N}\right) = 28\frac{4}{5}$.

In other words the ordinary lens must be placed in the position where the refraction caused by the Telephotographic lens *appears* to originate. (See Fig. 43.)

We thus confirm again our conclusion that a greater distance must intervene between object and lens with a Telephotographic lens than with an ordinary positive system, both having the same focal length, in order to produce the same size of image, or the same magnification.

If both lenses have the same intensity, or ratio of effective aperture to focal length, the Telephotographic lens has the advantage (1) in giving better perspective, and (2) in greater "depth of focus."

It must be remembered, however, that under these particular circumstances the Telephotographic lens will require a longer exposure than the ordinary lens in the proportion of the square of their distances from the object; in our particular case as $(174)^2 : (144)^2$, or very nearly as $3 : 2$.

Very little consideration will show the reason for this. When images of the same size are produced by either type of lens worked at the same intensity and at the same distance, the relative exposures must be the same; but here, although the size of the image is the same in both instances, the distance the light has to travel from the object before reaching the Telephotographic lens is greater, and the comparative illumination of the two images obeys the "law of inverse squares."

The reader will find more examples illustrating the use of the lens for photographing near objects in Chapter VII. on the "Application of the Lens."

To sum up, this method (B) seems to offer the following advantages to the photographer :

FORMATION OF ENLARGED IMAGES

(1) It only extends the already familiar principle of producing images of objects in a given scale.

(2) The magnification of the original scale carries with it direct information on the question of correct exposure.

(3) The determination of the "equivalent" lens of the system is arrived at more readily than in the method (A), when it is necessary to find it. In (A) it is *always* necessary, and involves a knowledge of the "interval" between the component lenses, as well as their focal lengths. In the method (B) the knowledge of the "equivalent" lens is only really necessary in the case of photographing near objects.

(4) Fewer preliminary considerations are required in arranging the scale for the final image. The distance between object and lens decides the scale given by the positive lens, and the subsequent magnification necessary is immediately known and attained.

(5) Less mental effort or calculation is required for equally accurate results.

CHAPTER VI

THE USE AND EFFECTS OF THE DIAPHRAGM, AND THE IMPROVED PERSPECTIVE RENDERING BY THE TELEPHOTOGRAPHIC LENS

IN determining the position and magnitude of the image of an object we have hitherto only made use of certain functions of a lens or lens-system, and, moreover, considered the latter as perfectly free from all aberrations, including distortion.

Reference to the four "cardinal" points is of great value and simplicity in this respect, but it is obvious that these geometrical constructions do not in reality trace the actual course of the rays which form the images. This is particularly obvious in reference to the Telephotographic construction. (See Fig. 42.) The rays utilised to determine the position, &c., of the image may not even pass through the lens at all.

It may be the mounting of the lens itself or the aperture of the diaphragm which limits and determines the portions of the lenses themselves that are utilised in the production of the various points forming the entire image. In Fig. 45 (1) shows that rays at a certain obliquity to the axis cannot pass the lens at all, but are intercepted by the mounting of the lens; (2) shows that the diaphragm with a given aperture in one position allows the entire pencil of rays at a given obliquity to pass through both lenses, a definite portion of each combination being utilised to form the image; (3) shows that if the same diaphragm be placed in a different position the same pencil of rays meeting the lens at the same obliquity as in (2) is almost entirely

USE AND EFFECTS OF DIAPHRAGM

cut off, but the figure shows that new portions of the positive and negative lenses are now utilised to form the image.

From diagrams (2) and (3) it is obvious that the size of the diaphragm will not only affect the different portions of the two lenses that are brought into play in forming the image, but also affect the illumination of the image, and the extreme obliquity at which a pencil of rays may pass through both lenses.

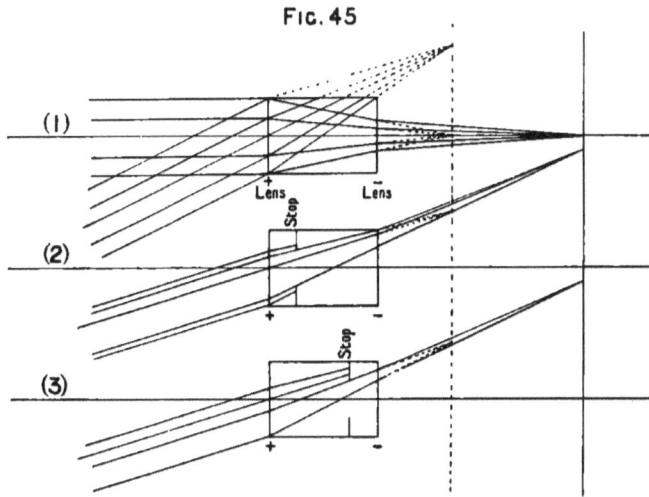

Fig. 45

A glance at Fig. 46 will show the position in which the diaphragm should be placed in order to transmit the greatest angle; this is seen to be the crossing point of those rays which traverse both lenses from their extreme edges. The diaphragm now also occupies the position for the greatest equality of illumination. If we project the extreme incident rays to meet in the axis, they form an angle a, which determines the extreme angle included when *no* diaphragm is employed.

TELEPHOTOGRAPHY

Let us now examine the effect of the diaphragm in greater detail.

On Rapidity.—In comparing the rapidity of two lenses, it is obvious that the one which gives an image of greater intensity, or possesses a higher degree of illumination than the other, is the more rapid of the two. In general we form an idea of rapidity of a lens by defining its *Intensity;* this we will denote by I.

The intensity is the ratio of the diameter of the effective aperture to the focal length of the lens. Calling a the diameter of the effective aperture and f the focal length,

$$I = \frac{a}{f} \quad \ldots \ldots \ldots \ldots (20)$$

It is important that we should clearly understand the meaning of

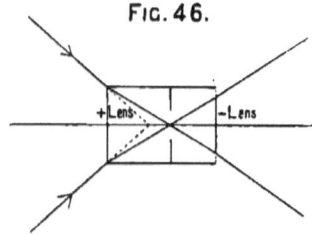

Fig. 46.

the "effective" aperture. If a beam of parallel rays (Fig. 47) is incident upon a positive lens A, having a certain focal length f, the rays converge after passing through the lens and meet in a point F, its focal point. The diameter of A A is the effective aperture of the lens, as none of the rays incident upon it are intercepted after convergence. If we place diaphragms B B, C C, of smaller diameter than A A, behind the lens as in (1), in positions which just allow the entire cone of rays A F A to pass, the diameters of B B and C C have the same effective apertures as A A. Calling a the diameter of A A, the intensity in each case is $\frac{a}{f}$. If, however, the diaphragm B′ B′ intercepts any of the rays transmitted

USE AND EFFECTS OF DIAPHRAGM

through A A as in (2) the effective aperture becomes then A' A'; calling b the diameter of A' A', then the intensity is $\dfrac{b}{f}$

For various reasons it may be necessary to reduce the diameter of the full effective aperture, and to accomplish this we make use of diaphragms whose effective apertures are arranged to have certain intensities. The intensities assigned to the diaphragms are indicative of the relative exposures necessary for each, and they are arranged so that the latter are easily comparable. The term "stop" is synonymous with "diaphragm."

As the Telephotographic combination is used to give a great variety of focal lengths, it is most convenient to adopt a system of

FIG. 47.

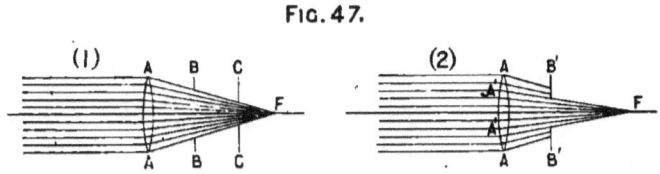

diaphragm notation that has direct reference to the focal length of its positive element.

The stops are denoted by the ratio of the focal length of the lens to the effective aperture. Thus if $\dfrac{f}{a} = 8$, then the stop is called $\dfrac{f}{8}$, or its intensity is $\dfrac{f}{8}$.

Fig. 48 shows the intensities (usually engraved on positive lenses) and relative rapidities of the diaphragm notation based upon the above system. In general, if the diameter of the aperture is known, with the focal length of the lens, and we express the intensity as above, the comparative exposures are as the squares of the denominators of the stop notation. Example: the relative exposures of stops $\dfrac{f}{10}$ and $\dfrac{f}{15}$ are as $10^2 : 15^2$, or as $1 : 2\tfrac{1}{4}$ (not given in the figure).

TELEPHOTOGRAPHY

In the last chapter we have considered the Telephotographic lens from two different standpoints; the method of treating it (B) will readily

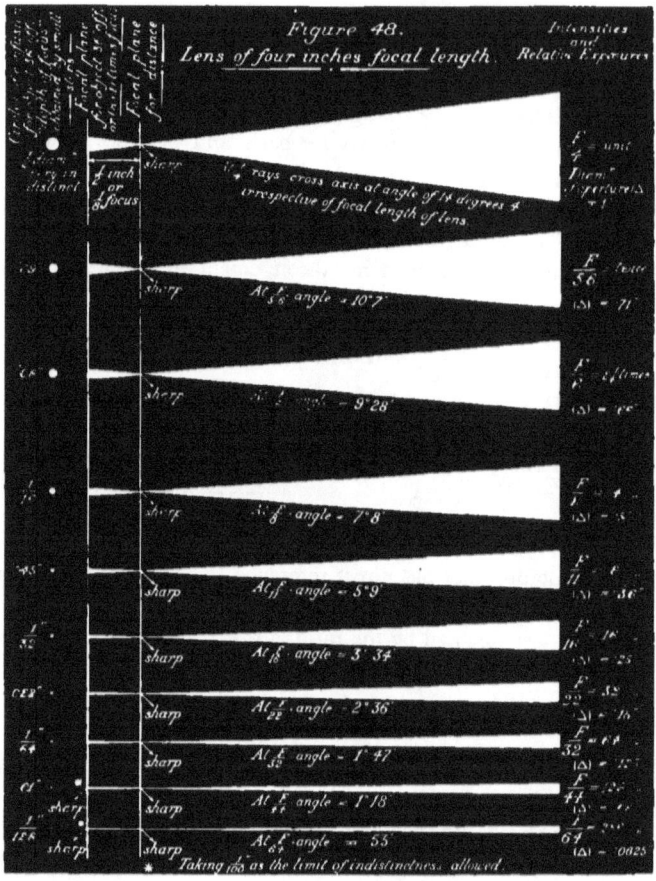

SCALE, 3/4.

commend itself for arriving at comparative exposures. It is only necessary to multiply the exposure required for the stop notation

84

USE AND EFFECTS OF DIAPHRAGM

indicated on the positive lens by the square of the magnification, and we know the exposure to be given for the Telephotographic lens. The knowledge of the effective aperture in some Telephotographic constructions is most important because the stop is sometimes considerably removed from the front converging lens. The absolute measurement of its diameter would be very different from its real effective aperture, and if used in place of the latter as a basis of calculation would lead to very great discrepancies, and consequent failure in the correct timing of exposures.

Fig. 48* exhibits at a glance the meaning of a "rapid" as distinguished from a "slow" lens. For convenience we have illustrated a lens of 4 inches focal length, but it is evident that the cones of light will be of the same shape, or the rays will cross at the same angle, for any given intensity, no matter what the focal length of the lens may be. The figure also shows that rapid lenses have less "depth of focus" than slow lenses, but we shall refer to this further on.

In designing a Telephotographic combination we must carefully consider the purpose for which it will be employed. If we take a positive lens of *high* intensity we can convert it into a Telephotographic system by combining it with a negative lens, which may be destined for either rapid work, or, on the other hand, to attain high magnification. For example, if we take a positive lens, having an intensity of $f/4$ we can employ a negative lens to magnify the image three times, and yet still have a fairly rapid lens in the combination, of intensity $f/12$. Again, if high magnification is the aim, we can magnify the image given by the positive lens as much as sixteen times and the combination still has a reasonable intensity of $f/64$.

On the other hand, if we take a positive lens of only moderate intensity, say $f/8$, and combine it with a negative lens, we can only expect to get small or moderate amplifications without necessitating a very low intensity in the Telephotographic combination.

The Lowest Intensity Permissible.—In treating of diffraction, Lord

* From "A Simple Guide to the Choice of a Lens." (J. H. Dallmeyer, Ltd.) By the Author.

TELEPHOTOGRAPHY

Rayleigh has demonstrated that the lowest intensity permissible without introducing the disturbing effects of diffraction is a ratio of aperture to focal length of 1 : 71, or $f/71$.

This conclusion has a very important bearing on the use of the Telephotographic lens. Suppose the intensity of the positive element is $f/8$ for example, and we wish to produce a magnification of five times, the combination will have an initial maximum intensity of $f/40$. We might very likely stop down the positive lens to $f/16$, or even less, with the object of getting finer definition. If the stop $f/16$ in the positive lens were employed, we should convert the intensity of the whole system into $f/80$, which is in reality too small for the finest definition.

We must then always remember that: the intensity of the positive element of the system divided by the magnification must not be less than $f/71$.

On Covering Power, or the size of plate covered at a given intensity. Referring to Fig. 49, the diameter of the circle beyond which no light can pass being called D, the effective aperture of the positive lens d_1 and the diameter of the negative lens d_2, and, as before, if f_1, f_2, represent their focal lengths, a the separation, and E the camera extension from the negative lens, it can be shown that:

$$D = E + f_2 \left\{ \frac{d_1 E f_2 + d_2 f_1 (E + f_2)}{f_2 [E (f_1 - f_2) + f_1 f_2]} \right\}$$

Approximate formula for maximum circle of illumination

$$= \frac{E}{f_2} \left(\frac{d_1 f_2 + d_2 f_1}{f_1 - f_2} \right) \quad \ldots \ldots \quad (21)$$

In the example shown in Fig. 43 $d_1 = 7/8''$, $d_2 = 5/4''$, $f_1 = 6''$, $f_2 = 3''$, and E = 9 inches. Hence

$$D = \tfrac{9}{3} \left(\frac{\tfrac{7}{8} \times 3 + \tfrac{5}{4} \times 6}{3} \right) = 10\tfrac{1}{8} \text{ inches.}$$

These formulæ only require a knowledge of arithmetic to arrive at the

USE AND EFFECTS OF DIAPHRAGM

result. The result by the approximation, $10\frac{1}{8}$, differs very slightly from that of the more complicated formula, which gives $10\frac{1}{10}$ as the diameter of the field.

If we employ a diaphragm and so reduce the effective aperture of d_1, we must substitute its value in the formula to find the circle of illumination covered, or the diameter of the field. Say its effective aperture is $\frac{1}{2}$ inch, then $D = 8$ inches.

In the formula it is evident that $d_1 f_2$ may become very small when the effective aperture d_1 is also very small.

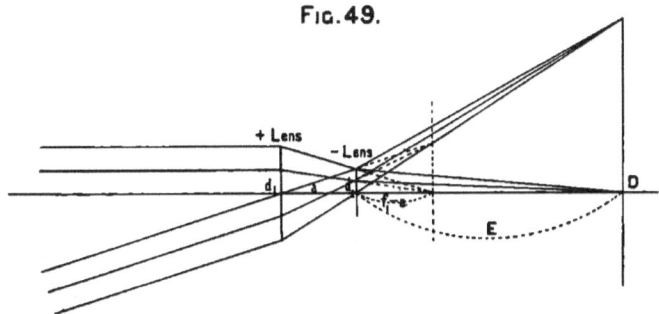

Fig. 49.

If we neglect this quantity we find the minimum circle of illumination for any lens

$$= (M - 1) \left(\frac{f_1 d_2}{f_1 - f_2}\right),$$

where M is the magnification as before.

In the case under notice

$$D \text{ min.} = (4 - 1) \frac{6 \times \frac{5}{4}}{6 - 3} = 7\frac{1}{2} \text{ inches.}$$

The circles of illumination must determine the size of the plate that can

TELEPHOTOGRAPHY

be covered; the diagonal of the plate must not be more than the diameter of the circle covered, or the corners of the plate will be "cut off," or not illuminated. It is advisable to arrange that the circle of illumination is greater than the diagonal of the plate for the smallest effective aperture likely to be used. We give here the diagonals of the sizes of plates in ordinary use.

Size of plate.	Diagonal or minimum diameter of circle of illumination.
$4\frac{1}{4} \times 3\frac{1}{4}$	5.4"
5×4	6.4
$6\frac{1}{2} \times 4\frac{3}{4}$	8
$8\frac{1}{2} \times 6\frac{1}{2}$	10.7
10×8	12.8
12×10	15.6
15×12	19.2
18×16	24.1
22×20	29.7
25×21	32.7
30×24	38.4

On the angle included by the combination.

Calling a the extreme angle, F the focal length of the Telephotographic lens, and D the diameter of the circle of illumination,

$$a = 2 \tan^{-1} \frac{D}{2F}.$$

In the case illustrated D=10.1 and F=24;

$$a = 2 \tan^{-1} \frac{10.1}{48} = 2 \tan^{-1} .210416.$$

From a table of natural tangents we find that .210416 corresponds to an angle of 11° 53′, and hence the extreme angle that can be included is 23° 46′. As the effective aperture of the diaphragm is decreased the diameter of the circle of illumination becomes less, or less angle is included as we stop down the lens.

PLATE XV Taken from a balloon from a height of 500 metres by Captain Mario Moris of the Italian Government, with a telephoto instrument specially designed for this purpose by the author.

USE AND EFFECTS OF DIAPHRAGM

It will be found, however, that when the lens is used to its utmost limits in respect of covering power, whatever stop is employed, the angle included is approximately a constant for any extension of camera ; if the lens is used to cover a definite size of plate at different extensions of camera, it is obvious that the angle included on that plate is variable, diminishing as we increase the magnification.

On Equality of Illumination.—Speaking of lenses generally, one form may lend itself more readily than another to transmit the full pencil of light over a greater angle, but no lens can in reality give equal illumination over the entire plate. This inequality is an inherent defect of every form of lens producing an image that is to be received upon a plane at right angles to the axis of the lens. It is most apparent when the combinations of the lens-system are separated by a considerable distance, particularly if used without a diaphragm. The construction of the Telephotographic lens necessitates a comparatively large separation of the elements composing it, even when m, or $\dfrac{f_1}{f_2}$ is small. Under these conditions the full pencil of rays from the front combination can only emerge from the back combination (without being cut off) over a very few degrees from the axis of the lens, or the centre of the plate. The tube of the lens then very rapidly commences to cut off the full pencil, and the worst conditions for equality of illumination are brought about.

A lens is readily examined to test the field over which the full pencil is received, by focusing upon a bright object, such as a candle or small gas flame, and racking the screen in a little, until the image becomes a bright disc of light. If we now move the camera so that the image gradually passes from the centre towards the edge of the screen, and observe how far the disc continues circular, we can follow how far the best conditions for equality of illumination are fulfilled ; as soon as the disc commences to become cut off, the full oblique pencil is no longer received upon the plate. If this operation is conducted, in the first place, without a diaphragm, we shall find that equality of illumination can be attained over a greater angle as the size of the diaphragm decreases, but at the expense of intensity.

TELEPHOTOGRAPHY

Even when the best conditions for equality of illumination are maintained, still the illumination of the plate rapidly decreases with the angle of obliquity. We state the case generally here as it is essential that we should choose our positive element, so that it, at any rate, shall transmit the full pencil of rays to the negative lens at the extreme obliquity.

Presuming that a full pencil passes through a lens at any obliquity of incidence, the quantity of light passing axially through the lens, as R R,

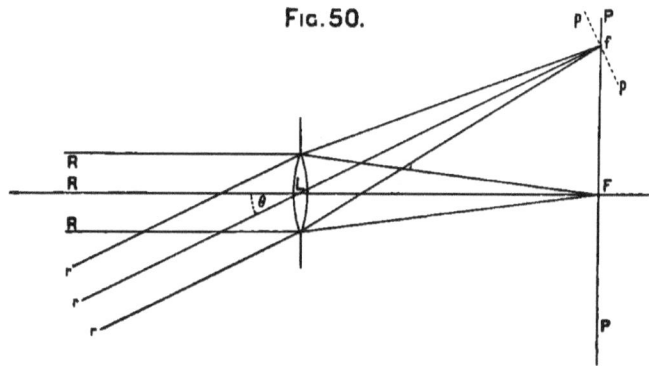

Fig. 50.

is greater than that which passes obliquely, the latter varying as the cosine of obliquity. Again the oblique pencils $r\,r$ are brought to a focus at f, more distant from the aperture of the lens L than the central pencil at F, the illumination varying according to the "law of inverse squares" when the plate is at right angles to rf, as at $p\,p$. Now L F and Lf vary with the secant of obliquity, and hence the illumination at F and f (on the plane $p\,p$) will vary as

$$\cos \theta \times \frac{1}{\sec^2 \theta} = \cos^3 \theta.$$

USE AND EFFECTS OF DIAPHRAGM

But the illumination on P P as compared with that of pp decreases again in the ratio of $\cos \theta$ (or as $\dfrac{1}{\sec \theta}$). Hence the final illumination varies as $\cos^4 \theta$, or as the fourth power of the cosine of the angle of obliquity.

The following table shows the illumination of the plate given by an ideally perfect lens as regards transmitting the full incident pencil at various obliquities.

Angle of obliquity $= \theta$.	Quantity of light passing through at the pupil $= \cos \theta$.	Illumination of image $= \cos^4 \theta$.
0°	1.000	1.000
5	.996	.985
10	.985	.941
15	.966	.870
20	.940	.780
25	.906	.675
30	.866	.562
40	.766	.344
45	.707	.250
50	.643	.171

Now, although this table shows that the illumination rapidly decreases for the greater angles of obliquity (one half when 60° and only one quarter when 90° are included upon the plate), the loss of light for small obliquities is seen to be very small up to 20° (that is 10° either side of the axis). The Telephotographic lens is seldom constructed to include a larger angle than this, so that we can readily choose a positive element which shall transmit light most favourably for equality of illumination ; but the separation of the elements, their diameters, and the position of the diaphragm must determine the limits for this condition in the compound system.

On Distortion.—If we place a diaphragm at any distance either before or behind a single positive lens, distortion of the image produced by it takes place, due to the fact that no single lens can be made free from spherical aberration, curvature of field, &c. ; a reference to

TELEPHOTOGRAPHY

Fig. 51 will make the matter clear: a beam of rays parallel to the axis of the lens L forms a focus at a point F in the axis; if the parallel beam $R_1 R R_2$ falls upon the lens obliquely, its focus (or more correctly its approximate focus) will be found at a point f which is not situated in the same plane as F, but if these rays are produced they will meet the focal plane of the lens through F in the points $r_1\ r r_2$. If we place a small diaphragm in contact with the lens, distortion will not take place, as both $r_1\ r_2$ are symmetrically situated with respect to the ray R r. If

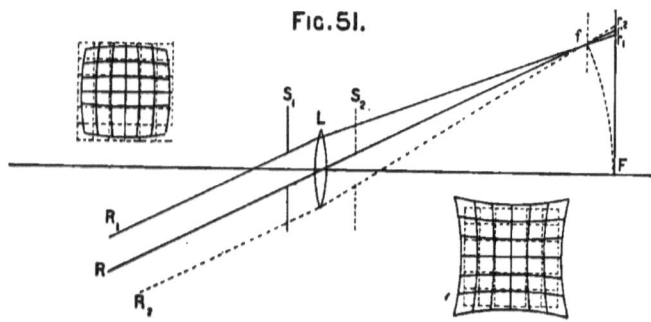

FIG. 51.

we remove the diaphragm from the position of contact, and place it in front of the lens in the position of S_1, the rays R_1 R only can traverse the upper portion of the lens L, and are received upon the plane through F in the points $r r_1$: the effect being to displace the ray $R_1\ r_1$ towards the centre of the image plane, the points which are most displaced being those furthest away from the centre; giving rise to what is known as "barrel shaped" distortion. If, on the other hand, we place the diaphragm in a position S_2 behind the lens, it will be seen that the rays R R_2 pass through the lower half of the lens L and are received upon the image plane in the points $r r_2$. The effect is to displace every point outwards from the centre, giving rise to what is known as "pincushion" distortion. These effects of distortion are best illustrated in the

PLATE XVI   Telephotograph of an Encampment taken from a balloon at a height of 800 metres by Captain Mario Moris.

USE AND EFFECTS OF DIAPHRAGM

rendering of a square or rectangular object as shown in the figure.

The figure has served to illustrate the manner in which distortion is produced by an uncorrected single lens; the same effects are noticeable in any single positive combination, and arise from the fact that, although adequate correction or defining power may be brought about in the axis of the system, it is impossible to remove spherical aberration, coma, curvature of field, &c., in the eccentrical pencils. These ill effects are reduced by employing a diaphragm which gives rise to the effects of distortion as indicated above.

The distortion of the image is in reality brought about by inherent defects in the lens itself, and not by the diaphragm employed; for example, if we place a diaphragm either before or behind an optical system which is perfectly corrected, such as a rapid rectilinear, stigmatic lens, &c., it will not bring about distortion in the image, and from this it will be obvious that were we enabled to produce a single combination absolutely free from all aberrations in the eccentrical pencils, the position of the diaphragm with regard to it would have no effect as regards distortion and have no influence in producing it.

From the above remarks it will be evident that the position of the diaphragm with respect to a single negative lens when placed in front of the latter will be to displace the pencils of rays falling upon it outwards from the centre. Hence, if a diaphragm be placed between a single positive and single negative lens, the effect will be to give "pincushion" distortion to the image formed by the positive lens alone, which in its turn will be emphasised by the negative lens in the final image. If we call $\frac{f_1}{f_2} = m$ as before, the "pincushion" distortion increases as m increases. A Telephotographic system, then, which has a single positive combination combined with a single negative combination with a diaphragm placed between must give this form of distortion in an inadmissible degree, even when m is low, except for pencils very near the axis, and in an impossible degree when m is great, or the lens is employed to cover a plate approaching the limits of its circle of illumination.

TELEPHOTOGRAPHY

If the positive element is a single combination with the diaphragm placed in front as in a "single landscape lens," which produces "barrel shaped" distortion *per se*, it can be combined with a single negative element, and form a Telephotographic system which is non-distorting. The chief drawback to this arrangement is the low intensity of the positive lens and consequently that of the entire system when fine definition is aimed at.*

The "pincushion" distortion involved in a Telephotographic lens composed of a single positive element of high intensity combined with a single negative lens soon led the author to abandon this form of lens, replacing the single front lens by a positive combination of high intensity free from distortion in itself, combining it with a negative combination constructed to give the minimum distortion even when m is very high.

The extra reflecting surfaces of the lenses thus combined constitute a theoretical disadvantage as regards the brilliancy of the image; but this is not found of practical moment. The more complicated system seems to warrant the expenditure of optical means to remove a palpable defect.

On the "Pupils" of a Lens-system.—Professor Abbé has defined these as extending the significance of the diaphragm or stop. We may consider each separate stop of a lens-system as forming an image by the lenses which are in front of it (that is towards the object). The stop whose image thus formed appears under the smallest angle from the object is defined as the "Aperture Stop," and its image is called the "Entrance Pupil" of the lens. Similarly the image of the "Aperture Stop" formed by the lenses succeeding it, that is towards the image, is termed the "Exit Pupil." The "Entrance Pupil" and the "Exit Pupil" bear to one another the same relation as object and image referred to the whole system. (See Notes.) These "Pupils" have a very important and interesting bearing upon the study of

---

* The author constructed a lens of this form for Mr. J. S. Bergheim to be used in portraiture. As Mr. Bergheim desired to produce "soft" images, the single positive element was made of high intensity, the spherical aberration necessarily introduced giving the softness aimed at, but the combination is free from distortion.

USE AND EFFECTS OF DIAPHRAGM

optical instruments. Von Rohr points out their bearing upon the correct method of viewing the perspective drawing given by a lens, and also upon the subject of "depth of focus," which is referred to below.

Perspective.—If we put a photographic lens in the orifice of a dark chamber so that it may form an image of any object situated outside, we can take a screen and place it at any distance we choose at (or beyond) the focal point of the lens, and thus receive upon it sharp images of distant objects or nearer objects according to the position in which the screen is placed. In other words, we can only have one plane in the field of the object, or the external field, which is strictly in focus at one time.

On the other hand, when we regard the external field, in order to form a correct idea of the perspective which will be produced in the resulting image upon the screen, we must first select the particular plane in the field of the object for which we shall focus sharply ; we must then consider the appearance of objects situated before or behind this particular plane as projected upon it; we then know, provided that the lens is non-distorting, that the perspective of the image will exactly reproduce the object plane and the projections of objects before or behind it, on to it, in some definite proportion. The perspective, then, is determined by the distance between the entrance pupil of the lens, through which pass all rays from the field of the object, and the chief plane for which we have focused. The entrance pupil and the exit pupil of the lens become the centres of perspective for object and image, and the image itself is an exact facsimile of the perspective produced upon the chief plane in the field of the object ; it is usually reduced in size in some proportion $n : 1$. Now, the size of the image itself does not give us any indication as to the correct point of sight, or distance from which it should be viewed. In order to discover this, we must place the image between the entrance pupil of the lens and the chief plane for which we have focused, in such a position that its projection exactly coincides with the objects in the object field.

In the case of employing a very small stop in the lens, all objects

TELEPHOTOGRAPHY

may, as a matter of fact, appear equally defined in the image, but as soon as we use a considerable aperture we see at once how objects lying either before or behind the selected object plane must become indistinct in the image, due to no inherent defect in the lens itself, but because when points in these objects are projected upon the chief plane of the object they are represented as circles, depending upon the diameter of the aperture of the lens, or the entrance pupil, and their distances from it. In general, every point in the object may be considered as sending out a cone of rays having the entrance pupil as a common base, every point in the object space forming a separate apex

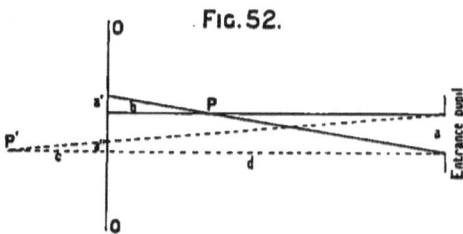

for each cone. Similarly the exit pupil forms a common base of a cone of rays for each point in the image space behind the lens.

If a represents the aperture of the entrance pupil, Fig. 52, and o o represents the plane for which we have focused and d the distance between them, we can form a true idea of the perspective given by the points P and P′ situated before and behind this plane respectively by considering P and P′ the apices of cones whose common base is the entrance pupil; these are seen to be a' and a'' respectively; a' and a'' are circles of indistinctness n times as large as the circles of indistinctness which will be found on the image plane when a reduction of n times takes place. This consideration leads to an interesting interpretation of the subject of "depth of focus," to which we shall shortly refer.

USE AND EFFECTS OF DIAPHRAGM

The correct distance from which an image formed by an ordinary photographic lens should be viewed is commonly defined to be a distance equal to that of the focal length of the lens. This rule is very approximately correct from the circumstance that the "Entrance" and "Exit" pupils of an ordinary positive lens very nearly coincide with the "principal" or "nodal" planes of the lens. When applied to the Telephotographic lens this condition is found not to hold. If we take a lens of ordinary construction, of given focal length, which reproduces objects in the chief plane reduced in a definite proportion $n : 1$, the object being a distance d from the lens, we find the correct distance x, Fig. 53, from which to view the image, thus :

$$d : x = n : 1,$$
$$\text{or } d = (n + 1)f,$$
$$\text{so that } x = \frac{n + 1}{n} f;$$

and if n be infinite $x = f$, which corresponds with the rule to view landscapes from a distance equal to the focal length of the lens above referred to. Now, if we take a Telephotographic lens of the same focal length, here

$$d = f(n + gm) ; g = 1, \text{ or is } > 1 ;$$

and the correct viewing distance

$$x = (\frac{n + g\,m}{n})f;$$

here, again, if n is infinite $x = f$.

These results have a very important bearing upon the perspective rendering of the Telephotographic lens.

When m and n are of the same order, which of course occurs in dealing with near and moderately near objects, x is considerably greater for this type of lens than is the case with an ordinary lens of the same focal length, the result being very closely connected with the fact

TELEPHOTOGRAPHY

that a greater distance is required between object and lens for the Telephotographic construction.

There are two errors which may be committed in looking at any photograph. It may be viewed from a point too far away, in which case the receding planes are dwarfed, and "distance" is exaggerated; or, on the other hand, if viewed from too near a standpoint, the opposite impression is conveyed, and the true sense of distance diminishes. Of these two errors the former is much more likely to occur, because, as already pointed out in Chapter II., the photographer is apt to approach his subject and include it under a large angle for the purpose of obtain-

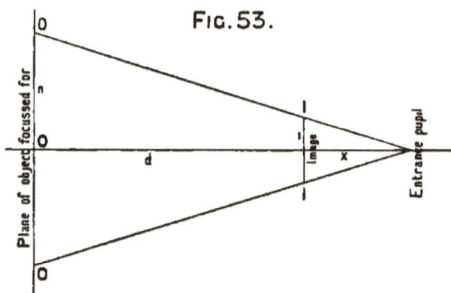

FIG. 53.

ing a sufficiently large image. The Telephotographic lens will enable him to remove the lens to a greater distance from the subject, and not only to include it under a considerably less angle, but to maintain the size of image that he desired with a lens of ordinary construction. To sum up, we find that with the Telephotographic lens we have the opportunity of employing a greater distance between the "Entrance Pupil" of the system and the chief object plane, thus including a less angle than would be possible in the case of an ordinary lens, and thereby obtaining more satisfactory perspective.

Depth of Focus (Von Rohr's interpretation).—The conception of the "Entrance Pupil" of a lens and its separation from the chief plane

USE AND EFFECTS OF DIAPHRAGM

of the image enables us to grasp very readily the meaning of this expression, and also the means of attaining it. Strictly speaking, there is no such condition as "depth of focus" in the image given by a photographic lens. We can only produce a perfectly sharp image of one plane of the object; points in the object situated on either side of it must be represented by circles of varying indistinctness. The size of these circles we can control to a considerable extent by aid of the diaphragm, or by reducing the size of the "Entrance Pupil." We have referred to each point of the object as forming a cone of rays having a common basis—namely, the "Entrance Pupil." Each point in the object is represented by a point somewhere in the image space, but only points in one plane of the object can be received as points in one plane of the image, all the apices of the cones which have a common basis in the "Exit Pupil" being now situated in this plane; all other points in the image space, or apices of the cones on the image side, being cut by this plane, form circles of indistinctness more or less extended. To attain so-called "depth of focus," we make use of a convention, and prescribe a limit to the size of these circles of indistinctness. As a matter of fact, this is somewhat difficult to do, or to state in a definite form, because the admissible degree of indistinctness will depend upon the distance from which it is viewed, particularly and more especially if the result is for pictorial effect and not for scientific accuracy. The convention usually adopted is founded on the fact that at the ordinary reading distance we cannot readily distinguish points which are less than one-hundredth of an inch apart, in other words, if a picture is made up of points which do not exceed circles of indistinctness greater than one-hundredth of an inch, it is said to appear "sharp."

In general, if we take i as the limit of indistinctness permissible in the image, and we reproduce the object in the proportion of $n : 1$, the circle of indistinctness for any point in the field of the object projected upon the object plane for which we have focused would be n times i, or $n\,i$. Referring to Fig. 52, the circles a' and a'' projected on to the chief plane of the object from the points P P' are, of course, n times as

TELEPHOTOGRAPHY

great as the circles of indistinctness to which they correspond in the image. Using the same notation as before, calling d the distance between the chief object plane and the "Entrance Pupil," b the distance of the point P in front of this plane, and c the distance of the point P' behind it, a the diameter of the "Entrance Pupil," and i the limit of indistinctness in the image,

$$b = \frac{d\,n\,i}{a + n\,i},$$
$$c = \frac{d\,n\,i}{a - n\,i}.$$

b is here the front depth of field, and c is the back depth of field, the whole depth being

$$b + c = \frac{2\,d\,n\,i\,a}{a^2 - n^2\,i^2}.$$

The size of the diverging circles a' and a'' is not only dependent upon the diameter of the "Entrance Pupil," but also on its separation from the chief plane for which we have focused.

Fig. 54 clearly shows that these circles decrease when the separation d increases. To put this more clearly, we see from the former figure that

$$a^1 = \frac{b\,a}{d - b};$$

now if we increase d by h the formula becomes

$$a^1 = \frac{b\,a}{d + h - b}.$$

If d is very great compared with $d-b$, or, in other words, if the point P and the principal object plane are both widely separated from the "Entrance Pupil," the effect of further removing the "Entrance Pupil" becomes indifferent. Thus we see the advantage gained by the Tele-

PLATE XVIII

"A photograph taken with Dallmeyer's telephoto lens of highest amplification during war at a distance of upwards of two miles. The negative by an official photographer of the Japanese War Department. The vessel is the *Tri-yen*, one of the two largest in the Chinese Navy, 7300 tons displacement, coated with steel 14 in. thick. Vessel half sunk, having been damaged by torpedoes. The effect of the cannon shot is noticeable in the white irregular lines just above the water mark." (Y. Isawa, *Editor of the "Shashin Sowa," Japan*.)

USE AND EFFECTS OF DIAPHRAGM

photographic lens in producing sharper or better defined images of near objects, giving the effect of greater depth of focus, and that the advantage ceases for very distant objects. The reason for this greater depth of focus in the Telephotographic lens lies in the smaller angles subtended by the "Entrance Pupil" from points in the object, as its distance is greater than for a lens of ordinary construction. Supposing the angle subtended is 2 u, then

$$\tan u = \frac{a}{2\,(d + h)}.$$

In conclusion we find that a Telephotographic lens of the same intensity as an ordinary positive lens gives greater depth of focus than the latter

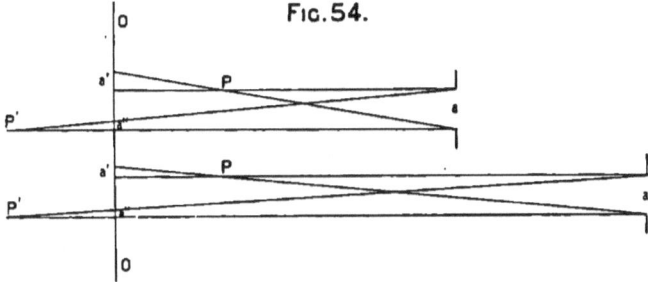

Fig. 54.

when used to photograph near objects, although this advantage gradually diminishes when the distance of the object increases, and is finally lost when the distance of the object becomes very great. It must be remembered, however, that the greater separation of the "Entrance Pupil" of the Telephotographic lens from the object, in circumstances where this advantage occurs, will necessitate an increase of exposure depending upon the "law of inverse squares" as already pointed out.

Depth of Focus (usual interpretation).—Our method of treating the Telephotographic lens, under the heading B, enables us again to

TELEPHOTOGRAPHY

simplify the question of depth of focus applied to the construction. If the positive element of the system be considered alone, we may apply the well-known formulæ for front and back "depth of field" in the usual manner, and if the result is to be regarded as a pictorial representation, we may neglect the effect produced by the magnification of the admissible circles of indistinctness which have been formed by the positive lens alone, because the more we magnify the primary image, the greater will be the distance from which the picture may be expected to be viewed. (This of course actually occurs in an ordinary "enlargement" from a photographic negative.) In selecting the circle of confusion of one-hundredth of an inch as sufficiently small when viewed at the normal distance of vision to appear practically sharp, we are guided by the fact that at a distance of about 12 inches this measurement subtends an angle of only three seconds of a degree. It is obvious that if we view the result obtained by the Telephotographic lens when a certain magnification has taken place, at a correspondingly greater distance than that at which we should view the result made by the positive lens alone, the increased circles of indistinctness will still only subtend the same angle at the eye, and therefore will *appear* equally well defined. If, on the other hand, we require the final image produced by the Telephotographic lens to have a *definite degree* of sharpness which shall not exceed one-hundredth of an inch, it will only be necessary to substitute the numeral M 100 (representing M times 100) for 100 in the formulæ, wherever 100 occurs, in order to attain the same limit in the final image, whatever M may be ; M as before being the magnification.

For the sake of uniformity and greater simplicity we here give formulæ for depth of focus which have reference to the relation between the size of object and image (or their distances from the focal planes of a positive lens), $\frac{1}{n}$ as before being "the magnification" of the image.

Calling f the focal length of the positive element, a its effective aperture, $I = \frac{a}{f}$ its intensity, $(n+1)f$ the distance of the object from the lens,

USE AND EFFECTS OF DIAPHRAGM

the distance beyond which all objects will be sufficiently well defined

$$= 100\, af + f \quad \ldots \ldots \ldots (22)$$

or 100 times the effective aperture multiplied by the focal length + the focal length of the lens ; or

$$= 100\, I f^2 + f \quad \ldots \ldots \ldots (22^a)$$

When we focus upon a near object the front depth

$$= \frac{nf(n+1)}{100\, a + (n+1)};$$

or

$$= \frac{n \text{ times the distance of the object}}{100 \text{ times the aperture } + (n+1)};$$

or

$$= \frac{nf(n+1)}{100\, I f + (n+1)} \quad . \quad . \quad (23)$$

The back depth

$$= \frac{nf(n+1)}{100\, a - (n+1)};$$

or

$$= \frac{n \text{ times the distance of the object}}{100 \text{ times the aperture } - (n+1)};$$

or

$$= \frac{nf(n+1)}{100\, I f - (n+1)} \quad (24).$$

Example.—An object is placed at a distance of 20 ft. from a Telephotographic lens composed of a positive lens f_1 of 10 inches focal length, working at an intensity $I = \frac{1}{4}$, and a negative lens f_2 of 5 inches focal length ; the primary image being magnified 4 times, or $M = 4$.

In order that the primary image shall not include circles of indistinctness greater than 1/100 we find, as $\dfrac{1}{n} = \dfrac{1}{23}$.

TELEPHOTOGRAPHY

Front depth

$$= \frac{23 f_1 (23 + 1)}{\frac{100}{4} f_1 + (23 + 1)} = 20 \text{ inches.}$$

Back depth

$$= \frac{33 f_1 (23 + 1)}{\frac{100}{4} f_1 - (23 + 1)} = 25 \text{ inches.}$$

Now, if we place the focusing screen at a distance of 15 inches from the negative lens, we shall obtain a magnification of four times. Should the result be regarded from the pictorial point of view, it will not be necessary to pay attention to the fact that we have magnified the primary image, for the reasons stated above.

But if we do not wish the limit of indistinctness to exceed $\frac{1}{100}''$ in the final image, then the front depth

$$= \frac{23 f_1 (23 + 1)}{\frac{400}{4} f_1 + (23 - 1)} = 5.4 \text{ inches.}$$

Back depth

$$= \frac{23 f_1 (23 + 1)}{\frac{400}{4} f_1 - (23 + 1)} = 5.7 \text{ inches.}$$

It will be observed that our process is to ascribe a smaller limit of indistinctness to the image produced by the positive lens proportionate to the amount it will be subsequently magnified.

We will now prove that of two lenses of the same focal length, one being of ordinary construction and the other of Telephotographic construction, that the latter has the greater depth of focus.

Example.—Let the focal length of both lenses be 30 inches, the Telephotographic combination being formed of a positive lens (*c. de v.* for instance) of 6 inches focal length and a negative lens of 3 inches

USE AND EFFECTS OF DIAPHRAGM

focal length. Let both have the same effective aperture of 2 inches, or I in each case be $\frac{1}{15}$. To find front and back depths for a "magnification" of $\frac{1}{4}$, or $n=4$

(1) For the ordinary lens:

Distance of object
$$= 150 \text{ inches.}$$

Distance of image
$$= 30 + 7\tfrac{1}{2} = 37\tfrac{1}{2} \text{ inches.}$$

Front depth
$$= \frac{4 \times 30 \times 5}{\frac{100}{15} \times 30 + 5} = 3'' \text{ (nearly).}$$

Back depth
$$= \frac{4 \times 30 \times 5}{\frac{100}{15} \times 30 - 5} = 3'' \text{ (slightly more).}$$

(2) For Telephotographic lens:

Distance of object
$$= 186 \text{ inches.}$$

Distance of image
$$= 12 + 7\tfrac{1}{2} = 19\tfrac{1}{2} \text{ inches.}$$

(12 inches extension are required to make the 6-inch positive lens equivalent to 30 inches focal length, by means of the 3-inch negative; the increased conjugate $7\tfrac{1}{2}$ inches is the same as in the case of the lens of ordinary construction.)

Here $n = \frac{186}{6} - 1 = 30$; and for $M = 7\tfrac{1}{2}$, $\frac{1}{100}$ becomes $\frac{1}{750}$, and $I = \frac{1}{8}$.

Front depth
$$= \frac{30 \times 6 \times 31}{\frac{750}{3} \times 6 + 31} = 3.64 \text{ inches.}$$

TELEPHOTOGRAPHY

Back depth

$$= \frac{30 \times 6 \times 31}{\frac{750}{3} \times 6 - 31} = 3.8 \text{ inches}$$

The following tables apply to measurements based upon an admissible circle of indistinctness of $\frac{1}{100}''$ in the image formed by the positive lens alone. For this degree of definition in the final image the distances given in the first table must be multiplied by the magnification given to the image.

TABLE I.

TABLE OF DISTANCES at and beyond which all objects are in focus and may be considered as situated in one plane.

*Intensities of Diaphragm Apertures.

Focal length of lens in inches.	F/4	F/5.6	F/6	F/7	F/8	F/10	F/11	F/15	F/16	F/20	F/22	F/32	F/44	F/64
	Number of feet distant after which all is in focus.													
4	33	24	22	19	17	13	12	9	8	7	6	4	3	2
4¼	38	27	25	21	19	15	14	10	10	8	7	5	3½	2½
4½	42	30	28	24	21	17	15	11	11	8½	7½	5½	4	3
4¾	47	34	31	27	24	19	17	12	12	9½	8½	6	5	3
5	52	36	35	30	26	21	19	14	13	10½	9½	6½	5½	3½
5¼	57	40	38	33	28	23	21	15	14	11½	10½	7	5½	3½
5½	63	45	43	36	31	25	23	17	15	12½	11½	7½	6	4
5¾	68	50	46	38	34	27	25	18	17	13½	13	8½	6½	4
6	75	54	50	42	38	30	28	20	19	15	14	9	7	4½
6¼	81	58	54	46	40	32	29	22	20	16	15	10	7½	5
6½	87	62	58	50	44	35	32	23	22	17½	16	11	8	5½
6¾	94	67	63	54	47	38	34	25	24	19	17	12	8½	6
7	101	72	68	58	51	40	37	27	25	20	18	12½	9	6
7¼	109	78	73	62	54	44	39	29	27	22	20	13½	10	6½
7½	117	83	78	64	58	47	42	31	29	24	21	14½	10½	7
7¾	124	90	83	71	62	50	45	33	31	25	22	15½	11	7½
8	132	96	88	76	68	52	48	36	32	28	24	16	12	8
8¼	141	100	94	80	71	56	51	37	35	29	25	17½	12½	8½
8½	150	104	100	84	76	60	56	40	38	30	27	19	13½	9
8¾	156	111	104	89	78	63	57	42	39	32	29	20	14	10
9	168	120	112	96	84	67	61	45	42	34	31	21	15	10½
9¼	180	127	116	101	90	71	65	47	45	35	32	22	16	11
9½	190	133	125	107	95	74	68	50	47	37	34	24	17	12
9¾	197	141	131	113	99	79	72	52	50	39	36	25	18	12½
10	208	148	140	120	104	83	75	55	52	42	38	26	19	13

* Intensities marked white on black ground are illustrated in Fig. 48.

USE AND EFFECTS OF DIAPHRAGM

In the annexed table the front and back depth must be divided by the magnification given to the positive lens, if the same degree of definition is required in the final image.

As the Telephotographic lens may be used with advantage in the production of direct *enlarged* images of objects which are situated approximately (or entirely) in one plane, we have included quite small multiples of the focal lengths of the positive lenses in the above table. These of course correspond to distances which are far too small for rendering good perspective, or the pictorial representation of objects lying in different planes.

The application of Table I. is obvious.

The application of Table II. is as follows : First select the standpoint from which the amount of subject to be included upon the plate *appears* in good perspective to the eye. Suppose the distance between lens and subject, in a particular case, to be 10 ft. Now the focal length of the positive element of the Telephotographic lens determines the *scale* of the primary image at this particular distance, and its intensity controls the front and back depth. Say the positive element is a "Cabinet lens" of 12 inches focal length; the scale is then $\frac{1}{9}$, or 18 inches in the length of the subject will occupy 2 inches in the primary image. Beneath the column " 10 feet " in the table, and opposite the lens of 12 inches focal length, we find both the scale of reproduction and the front and back depth for a limit of indistinctness of $\frac{1}{100}$ of an inch. We must regulate the intensity by the amount of depth required. Say this is 10 inches ; we find that at an intensity of $f/6$, that the combined front and back depth is rather more than this. Hence at a distance of 10 ft. the positive element of our combination, which has a focal length of 12 inches, gives an image $\frac{1}{9}$ the size of the object and that a diaphragm of intensity $f/6$ must be used for the required definition throughout the image.

We may now proceed to consider the magnification to be given by the negative element with which it is combined to form the Telephotographic system. Let the negative element have a focal length of 6 inches, and suppose we wish to reproduce the object finally as half

TELEPHOTOGRAPHY

size, or in the scale $\frac{1}{2}$. We must magnify the primary image $4\frac{1}{2}$ times; to do this we set the screen $3\frac{1}{2}$ times the focal length of the negative lens from it, or at a distance of 21 inches. An adjustment between the separation of the two elements will enable us to focus accurately upon the screen in this position.

Any departure from the finest possible definition given by the positive lens (not exceeding $\frac{1}{100}$ of an inch) will be magnified $3\frac{1}{2}$ times in the final image. This final degree of definition may or may not suffice, according to the requirements of the case. For a large portrait it will probably be sufficiently well defined, as the image will not be viewed too closely. On the other hand, if an absolute degree of definition is required through the depth of the image, equal to that in the primary image for example, it will be necessary to select an intensity which gives $4\frac{1}{2}$ times the depth necessary for the positive element. This is roughly given by dividing the intensity $f/6$ by the magnification to which the primary image has been subjected.

NOTES.

On the "Pupils" of a Lens-system.—Let us first see how to construct the images of a diaphragm which are formed by the separate portions of a lens-system. In Fig. 55 the upper diagram consists of a lens-system composed of two positive lenses, L_1, L_2 respectively, the stop s being situated midway between them. In order to find the image of the stop formed by the lens L we draw a line parallel to the axis meeting the lens L_1, which after refraction must pass through its focal point f_1; again a ray from the edge of the stop s passing through the centre of the lens L_1 continues in a straight line; it will be seen that they meet in a point s_1; we thus find the image of any stop s formed by the lens L_1 at s_1. Similarly we determine the position of the image of the stop s formed by the component element of the system L_2 situated behind the stop, and find its image in s_r.

In the lower figure the front element of the system is a positive lens, and the back element is a negative lens; we construct the image

USE AND EFFECTS OF DIAPHRAGM

of the stop s formed by the lens L_1 and find it at s_1; and using the same construction as before for determining the image of the stop s by the negative lens L_2, we find its image at s_2.

It will be observed that in all cases the stop images in these constructions are virtual images. It is evident that s_2 is the image of s_1 formed by the whole lens-system, and s_1 and s_2 are therefore conjugate to one another; so that any ray which passes through the image s_2 must before refraction necessarily have passed through s_1, and hence a

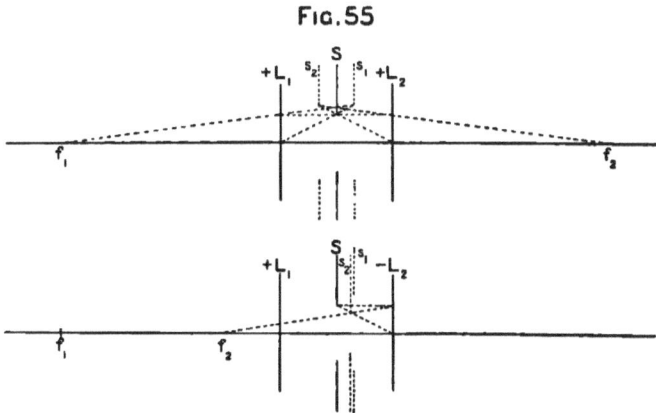

Fig. 55

stop or diaphragm in the position and of the size s_1 will have the same practical effect in limiting the pencils which form the image as a stop in the position and of the size of s_2. Both have the same effect as the actual stop s.*

When we have determined the position of these stop images we are enabled to trace the actual course of the rays which pass through any lens-system by an accurate and pretty geometrical construction, and thus determine the image of any object. Referring to Fig. 55A, if

---

* Dr. S. Czapski, "Theorie der optischen instrumente nach Abbe," pp. 155, 156.

TELEPHOTOGRAPHY

TABLE
TABLE OF FRONT AND
FOR AN OBJECT AT

DISTANCE OF OBJECT.		3 FEET.			4 FEET.			5 FEET.			6 FEET.			8 FEET.		
Focal length of Lens in inches.	Intensity I.	Scale, or $\frac{1}{n}$	Front depth in inches.	Back depth in inches.	Scale, or $\frac{1}{n}$	Front depth in inches.	Back depth in inches.	Scale, or $\frac{1}{n}$	Front depth in inches	Back depth in inches	Scale, or $\frac{1}{n}$	Front depth in inches.	Back depth in inches.	Scale, or $\frac{1}{n}$	Front depth in inches.	Back depth in inches.
*6"	$\frac{F}{3}$	$\frac{1}{5}$	.9	.93	$\frac{1}{7}$	1.61	1.75	$\frac{1}{8}$	2.05	2.26	$\frac{1}{11}$	3.73	4.21	$\frac{1}{15}$	6½	7¾
	$\frac{F}{4}$		1.15	1.25		2.13	2.36		2.71	3.06		4.8	5½		8½	10⅔
	$\frac{F}{5}$		1.43	1.6		2.62	3		3.35	3.89		6	7¼		10½	13¾
	$\frac{F}{6}$		1.7	1.9		3.1	3.65		3.96	4.76		7	9		12½	17¼
	$\frac{F}{8}$		2.22	2.6		4.05	5.01		5¼	6½		9	12½		15⅞	24½
	$\frac{F}{10}$		2.7	3.33		4.94	6½		6¼	8½		11	16½		19	32¾
*8¼	$\frac{F}{3}$	$\frac{1}{3.3}$	.42	.43	$\frac{1}{4.8}$	.81	.85	$\frac{1}{6.3}$	1.34	1.42	$\frac{1}{8}$	2.09	2.23	$\frac{1}{11}$	3.79	4.14
	$\frac{F}{4}$		.55	.58		1.08	1.14		1.77	1.9		2.75	3.01		4.99	5½
	$\frac{F}{5}$		.69	.73		1.34	1.44		2.2	2.4		3.41	3.8		6¼	7
	$\frac{F}{6}$		.82	.87		1.6	1.74		2.62	2.91		4.05	4.62		7¼	8¾
	$\frac{F}{8}$		1.09	1.24		2.1	2.4		3.43	3.96		5¼	6¼		9½	12
	$\frac{F}{10}$		1.35	1.49		2.6	2.99		4.22	5.04		6½	8		11½	15½
*10"	$\frac{F}{3}$	$\frac{1}{2.6}$	.28	.29	$\frac{1}{3.8}$	.54	.55	$\frac{1}{5}$	.88	.91	$\frac{1}{6.2}$	1.31	1.37	$\frac{1}{8.6}$	2.41	2.55
	$\frac{F}{4}$		.36	.38		.71	.74		1.17	1.22		1.73	1.83		3.18	3.43
	$\frac{F}{5}$		.45	.47		.89	.93		1.45	1.54		2.15	2.31		3.94	4.33
	$\frac{F}{6}$		.55	.57		1.06	1.12		1.73	1.86		2.56	2.8		4.68	5.26
	$\frac{F}{8}$		.72	.77		1.4	1.51		2.28	2.52		3.37	3.79		6¼	7¼
	$\frac{F}{10}$		.9	.97		1.74	1.91		2.83	3.19		4.16	4.81		7½	9¼
*12"	$\frac{F}{3}$	$\frac{1}{2}$	.17	.18	$\frac{1}{3}$	.35	.36	$\frac{1}{4}$	.59	.6	$\frac{1}{5}$	.88	.91	$\frac{1}{7}$	1.64	1.71
	$\frac{F}{4}$		.23	.24		.47	.48		.78	.81		1.17	1.22		2.18	2.3
	$\frac{F}{5}$		.29	.3		.59	.6		.98	1.02		1.46	1.54		2.71	2.85
	$\frac{F}{6}$		.35	.36		.7	.73		1.17	1.23		1.74	1.85		3.23	3.5
	$\frac{F}{8}$		.47	.49		.93	.98		1.55	1.65		2.31	2.5		4.26	4.73
	$\frac{F}{10}$		.58	.61		1.16	1.24		1.92	2.08		2.85	3.15		5¼	6

* These particular focal lengths are chosen as they correspond to those of well known into Telephotographic instruments

110

USE AND EFFECTS OF DIAPHRAGM

II

BACK DEPTH OF FOCUS

GIVEN DISTANCES

10 FEET.			12 FEET.			14 FEET.			16 FEET.			18 FEET.			20 FEET.			21 FEET.		
Scale, or $\frac{1}{n}$.	Front depth in inches.	Back depth in inches.	Scale, or $\frac{1}{n}$.	Front depth in inches.	Back depth in inches.	Scale, or $\frac{1}{n}$.	Front depth in inches.	Back depth in inches.	Scale, or $\frac{1}{n}$.	Front depth in inches.	Back depth in inches.	Scale, or $\frac{1}{n}$.	Front depth in inches.	Back depth in inches.	Scale, or $\frac{1}{n}$.	Front depth in inches.	Back depth in inches.	Scale, or $\frac{1}{n}$.	Front depth in inches.	Back depth in inches.
$\frac{1}{19}$	10⅓	12¾	$\frac{1}{23}$	14¾	18¾	$\frac{1}{27}$	20	26¼	$\frac{1}{31}$	25¾	35¼	$\frac{1}{35}$	32	46	$\frac{1}{39}$	39	58½	$\frac{1}{47}$	54½	89
	13½	17½		19	26¼		25½	37¼		32¾	50½		40¾	66¼		49¼	85		68½	132¾
	16¼	22¾		23	34½		30¾	49¼		39¼	67½		48½	90		58½	117		80½	188
	19	28⅓		26¾	43½		35½	63		45¼	87½		55½	118		67	156		91½	260¼
	24	41⅓		33⅓	65		44	96½		55½	138½		68	194		81½	267½		110	501⅓
	28½	57		39½	92		51¼	141¾		64⅝	212½		78¾	315		93½	468		125⅓	1128
$\frac{1}{14}$	6	6¾	$\frac{1}{17}$	8½	9¾	$\frac{1}{20}$	11¾	13⅝	$\frac{1}{23}$	15¼	18	$\frac{1}{26}$	19¼	23½	$\frac{1}{29}$	23½	29¼	$\frac{1}{35}$	33½	43½
	7¾	9		11¼	13½		15¼	18¾		19¾	25		24¾	32⅓		30¼	40¾		43	61
	9¾	11½		13¾	17½		18⅝	24		24	32¼		30¼	42		36¾	53¼		51¾	80½
	11¼	14⅛		16¼	21		21¾	30		28¼	40⅛		35¼	52½		42¾	66¾		60	102½
	14½	19½		20¾	29¾		28	42¼		35⅞	57½		44½	76		54	98¼		74¾	154¾
	17¾	25¾		25	39		33½	56½		42¾	78		53	104¼		63¾	136¾		87¾	223½
$\frac{1}{11}$	3.82	4.11	$\frac{1}{13.4}$	5⅓	6	$\frac{1}{15}$	7	7⅓	$\frac{1}{18}$	9¾	11	$\frac{1}{20.5}$	12⅓	14¼	$\frac{1}{23}$	15⅓	17¾	$\frac{1}{29}$	24	28¾
	5.04	5.54		7¼	8¼		9	10¼		12¾	14¾		16¼	19¼		20	24½		31	39½
	6¼	7		9	10¼		11	13		15½	19		20	24¾		24¾	31¼		38	51¼
	7½	8½		10⅝	12⅝		13	16		18½	23¼		23½	30¼		29	38¾		44¼	63½
	9½	11¾		13¾	17½		17	22		23¾	32¼		30	42½		37	54½		56	91½
	11¾	15		16¾	22½		20¾	28⅓		28¾	42¼		36¼	56		44½	72½		67	124½
$\frac{1}{9}$	2.63	2.77	$\frac{1}{11}$	3.84	4.08	$\frac{1}{13}$	5¼	5½	$\frac{1}{15}$	7	7⅓	$\frac{1}{17}$	8¾	9½	$\frac{1}{19}$	11	12	$\frac{1}{23}$	15½	17⅓
	3.48	3.72		5	5½		6¾	7½		9	10		11⅓	13		14¼	16¼		20½	24
	4.32	4.79		6¼	7		8½	9½		11¼	12¾		14¼	16½		17½	20¾		25	30½
	5.14	5.69		7½	8⅓		10¼	11½		13¼	15¾		16¾	20		20¾	25¼		29⅓	37⅛
	6¾	7¾		9¾	11⅓		13	15[illegible]		17¾	21½		21¾	27¾		26¾	35		38	52½
	8¼	9¾		12	14½		16	20¼		21¼	27¾		26½	36		32½	45⅓		46	69

Carte de Visite and Cabinet lenses of high intensity, and are well suited for conversion for taking *large* portraits in the studio.

Fig. 55.A

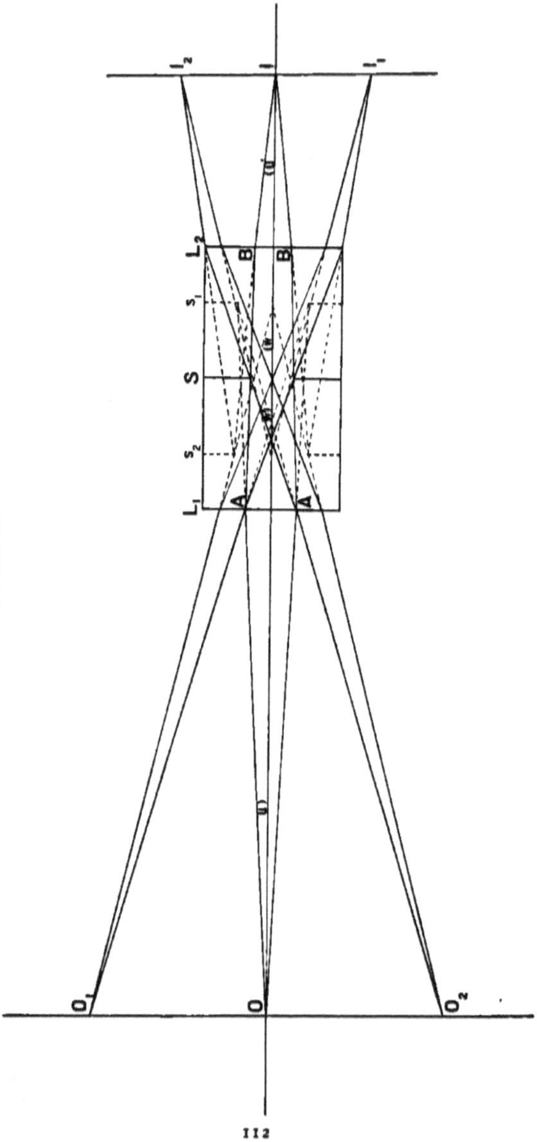

USE AND EFFECTS OF DIAPHRAGM

from a point o in the object plane o_1 oo_2 we draw a pencil of rays to the edge of the image of the stop s formed by the front lens at s_1, it will be found to cut the lens L_1 at the points A A; the actual course of the rays, after passing this lens, will be through the edge of the stop s itself, in the direction A B, A B meeting the lens L_2 at B B. These rays emerge from the lens L_2 as though they had proceeded from the image of the stop s formed by the back lens at s_2, and meet in the point I of the image. Professor Abbé defines the image of the stop s formed at s_1 which appears from o under the smallest angle (2 u) as the "entrance pupil" (eintritts pupille) of the system; similarly the stop whose image s_2 appears from I under the smallest angle (2 u) he terms the (austritts pupille) "exit pupil."

In a similar manner to that employed for tracing the course of the pencil of rays from o through the entire system, we may construct the image formed by any points o_1 o_2 in the plane of the object forming images at I_1 I_2 respectively; the dotted lines in the figure indicate the construction. It is evident that o_1 and o_2 may occupy positions such that rays proceeding from objects further from the axis cannot emerge from the lens at all; when o_1 and o_2 are situated in such positions that rays proceeding from them can only just emerge through the entire system, o_1 o_2 will appear from the centre of the "entrance pupil" s_1 under an angle 2 w; 2 w determines *the angle of the field* included by the lens. The image I_1 I_2 forms an angle 2 w_1 from the centre of the "exit pupil" s_2, and appears from it under the smallest angle.

We thus find that all the pencils of a system are of two kinds: (1) those which have their common base in one of the "pupils" and their apices in either the field of the object or the field of the image; (2) those which have their common bases in either the object or the image, and their apices in the corresponding "pupil."

CHAPTER VII

PRACTICAL APPLICATIONS OF THE TELEPHOTOGRAPHIC LENS

I.—IN PORTRAITURE

IT is probably a matter of common observation that the finest portraits are produced by photographers having the advantage of a long studio.

Where the studio is sufficiently long, the photographer is always able to take up a standpoint from which a study of a head, head and shoulders, and so on to full length of the figure, is *seen* to the best advantage. In general, the distance must be increased as the amount and depth of subject which he intends to portray increases.

In order to do himself and his subject justice, these *distances* should never be departed from, whatever the scale of the image is to be. The lens should be placed at the correct viewing distance, and the focal length of the lens chosen to give the scale required, determining the size of the final image.

No one positive lens of ordinary construction and definite focal length can answer all purposes satisfactorily. If the focal length be great, suitable for a "life-size" head for example, the lens would be quite unsuitable for taking a "Cabinet" standing figure, because it would have to be removed so far from the subject that the image would appear flat. Conversely, if a lens of shorter focal length, suitable for the "Cabinet" standing figure, were employed to take a "life-size" head, it would have to be brought so close to the sitter that the drawing of the head would be altogether unsatisfactory.

The general tendency in ordinary photographic practice is to

PRACTICAL APPLICATIONS OF LENS

commit the latter error—viz., to aim at too large an image with a lens which is too short in focal length to allow an adequate distance to intervene between the subject and lens for good perspective rendering.

The Telephotographic lens affords the great advantage of always enabling the larger sizes of portraits to be taken under favourable conditions for proper perspective rendering. Again, its perspective is always better than that given by an ordinary positive system of the same focal length—due to the greater distance that may intervene between subject and lens.

Let us now illustrate some comparative examples of the employment of ordinary and Telephotographic constructions.

The production of a "life-size" head : We know that to accomplish this with an ordinary lens the sitter must be at a distance equal to twice the focal length of the lens from it. Suppose we take a lens of 5 inches diameter and 40 inches focal length (intensity $\frac{F}{8}$); the distance of the sitter from the lens must be 80 inches, and the camera extension also 80 inches. A distance of 80 inches is none too far for good perspective rendering, and yet it is seldom that even this distance intervenes between lens and sitter for a "life-size" head, because the focal lengths of most large portrait combinations are less than half this distance, or 40 inches.

Let us now form a Telephotographic lens by combining a rapid cabinet lens of $3\frac{1}{2}$ inches diameter and 10 inches focal length with a negative lens of 4 inches focal length, as illustrated in Fig. 56.

If we place the combination at the same distance (80 inches) from the sitter, the positive element alone gives an image $\frac{1}{7}$ the size of the original. In order that the final image may be of the same size as the original, we must magnify it seven times. To do so, the focusing screen must be placed at a distance E from the negative lens, where $E = f_1 (M-1) = 4 (7-1) = 24$ inches, as compared with 80 inches above, see (13). On adjusting the separation between the lenses by means of the rack and pinion mount, the image will now be found to be of the same size as the original. To compare our present conditions for

TELEPHOTOGRAPHY

rapidity with those above, we must determine the focal length of the combination when thus employed. Substituting in (18) p. 75 where $E=24$, $m=\dfrac{10}{4}=2\tfrac{1}{2}$, $f_1=10$ and N (for equal size)$=1$;

Focal length
$$= \dfrac{m E + f_1}{\dfrac{m}{N} + 1}$$
$$= \dfrac{60 + 10}{2\tfrac{1}{2} + 1} = 20 \text{ inches only,}$$

and hence this combination will work at the same initial intensity as the former with an aperture of only $2\tfrac{1}{2}$ inches.

Fig. 56.

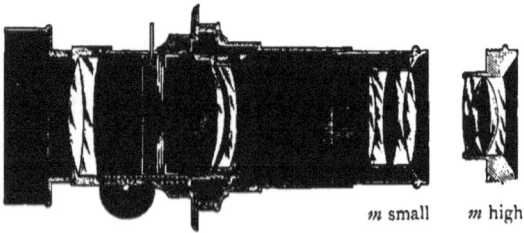

m small m high

Observe that as the lens produces the image of equal size with the object, the position of the screen must be 20 inches distant from its focal point, or focus for parallel rays. In other words, the combination focused for a very distant object would form an image at a distance of 24−20, or 4 inches from the negative lens. For parallel rays we see from (13) $F = m\ E + f_1 = 10 + 10 = 20$ inches, and our conclusion is confirmed.

From this result we can proceed to emphasise the valuable and important fact that a greater distance must intervene between a Telephotographic lens and the subject than is required for an ordinary

PLATE XXI

Façade, San Michele, Lucca. Taken with an ordinary lens 12-in. focal length. (*Copyright of and kindly lent by J. W. Cruickshank, Esq.*)

PLATE XXII

South-West Corner, Second and Third Stories of Façade, San Michele, Lucca. Taken with 6-in. Ross Universal Symmetrical with 3-in. negative. 18¾ in. in draw from back of lens to plate. Stopped down to $\frac{f}{16}$. 56 seconds exposure. (*Copyright of and kindly lent by J. W. Cruickshank, Esq.*)

PRACTICAL APPLICATIONS OF LENS

lens of the same focal length. With the former we have found 80 inches is required, while with the latter it must be twice 20, or 40 inches. These results also exemplify in a striking manner the fallacy of the usual tenet that the correct viewing distance of the photograph should be determined by the focal length of the lens with which it was taken. The result by the 20-inch Telephotographic lens and the 40-inch lens of ordinary construction should of course be viewed from the same standpoint when the result is "life-size"; and, again, we see that if both lenses have the same focal length the result produced by a lens of ordinary construction should be viewed at a distance of 40 inches only as against 80 inches by the Telephotographic system. The latter is the more likely distance from which a "life-size" head would be viewed, and the result by the Telephotographic lens will appear in better drawing and be a more truthful representation.

Although there is a considerable gain in using the Telephotographic lens, both as regards improved perspective and greater depth of focus, as compared with the lens of ordinary construction, there is a loss in rapidity when two lenses of the *same focal length* are compared. The comparative rapidities are now influenced by the "law of inverse squares" and their exposures are proportional to the squares of their respective distances from the subject. We have seen, however, that a Telephotographic lens of considerably shorter focal length than an ordinary lens can produce the same size of image at an *equal distance*, and hence the diameter of the positive element in the Telephotographic system may be smaller and yet give the same rapidity; this will be found both an advantage and an economy.

Reverting to the case of a "life-size" representation of a head, it is obvious that the distance of 80 inches is not by any means binding in the case of one and the same Telephotographic combination. Both theoretically and practically this lens may be placed at any greater distance we choose from the sitter. If a greater distance be chosen the scale of the image produced by its positive element becomes smaller and therefore requires greater magnification by its negative element in order to yield the same size of image finally.

TELEPHOTOGRAPHY

It is probable that a distance of 10 or 12 ft. will be chosen for the representation of a "life-size" head. At 10 ft., employing a Telephotographic lens composed of a cabinet lens of say 12 inches focal length and 4 inches aperture as positive element, and a negative lens of 6 inches focal length as negative element (to vary the numerical examples), the primary image will be $\frac{1}{9}$ of the original, and we must magnify it nine times for equal size. This requires a camera extension of 6 (9−1) = 48 inches, which shows from (18) that the lens has a focal length of 36 inches, and therefore an initial intensity of $f/9$.

If we wish to produce the image in the scale of half natural size, it is evident that all we have to do is to magnify the image $4\frac{1}{2}$ times. In this case we must take the camera extension 6 ($4\frac{1}{2}$−1) or 21 inches. The focal length of the system is now 26 inches, and the intensity
$$\frac{4}{26} = \frac{F}{6.5}.$$

It is hardly possible to give any hard and fast rules for the method of procedure in the various requirements of large portraiture, but it may safely be said that if a certain carte de viste or cabinet lens produces the size of image for which it was constructed at a given distance from the sitter, in really satisfactory perspective, these images may be taken as the basis of subsequent magnification by the negative elements with which they are combined to form Telephotographic systems. It is hoped that Table II. (p. 110) may be of considerable assistance in this respect.

The magnification of the primary image produced by the positive element can be accomplished in either of two ways : (1) By using a negative element of very short focal length in comparison to the positive element, or, in other words, making $\frac{f_1}{f_2}$, or m, great, and using only a very small camera extension ; or (2) by using a negative lens of longer focal length, or making m small, and obtaining the same magnification by the necessarily greater camera extension. The latter method is far preferable in portraiture, and, in fact, in almost all cases, as the conditions favour greater covering power, more even illumination, and freedom from distortion of the lines near the margin. For

PRACTICAL APPLICATIONS OF LENS

portraiture the focal length of the positive element should not be more than 2 to $2\frac{1}{2}$ times as great as that of the negative lens, or m should not exceed $2\frac{1}{2}$.

Plates I. and II. show a nearly full-length portrait of a child; Plate I. was taken at a distance of 10 ft. with an ordinary cabinet lens constructed for a *short* studio; Plate II. was taken by a *c. de v.* lens of $8\frac{1}{4}$ inches focal length, combined with a negative lens of about half its focal length, at a distance of 24 ft. The difference in the drawing is apparent; in Plate I. the hand and flower are too large, the background falls away too rapidly and appears too small, and the pedestal is out of drawing; in Plate II. the relative planes of the picture take their proper proportions. Plate III. illustrates a cabinet head, taken by the same Telephotographic combination at a distance of 15 ft. Again Plate IV. was also taken by the same Telephotographic combination at a distance of 10 ft. (N.B. It could have been as *sharp* as desired; the method of producing the soft effect is referred to later), and Plate V. was taken by a skilful professional photographer, Mr. Habgood of Boscombe, who was asked to produce a head of the same size by his ordinary methods. He naturally selected the lens of the greatest focal length in his possession (a rapid rectilinear of 16 inches focal length) for the purpose. The distortion of the features in this presentment, as the lens had to be brought within 4 ft. of the sitter, is apparent; the face appearing "bulged" and the size of the mouth distinctly exaggerated. Both negatives were purposely left untouched.

In addition to the precautions necessary for giving good *perspective* rendering in the image, the ill-effects of distortion in the lens itself must be carefully guarded against. For this reason the author gives preference to a non-distorting, well-corrected compound positive element, as against a single cemented lens of high intensity. Messrs. Zeiss have recently introduced a very interesting single cemented lens, illustrated in Fig. 57, which has the same high intensity as a portrait combination $f/3$. This is designed for combination with a single cemented negative lens illustrated in the same figure, in which m is advisedly kept small (as in such a combination, where the diaphragm is

TELEPHOTOGRAPHY

placed between the combinations, the distortion increases as m increases), and the photographer is warned not to utilise its covering power to the limits on account of the unavoidable distortion produced at and near the margins. Messrs. Zeiss describe the practical working of this lens, as they do that of all their Telephotographic lenses, in accordance with the method of treating the system under heading A., Chapter V. Dr. Rudolph, in his "Monograph" already referred to, furnishes a

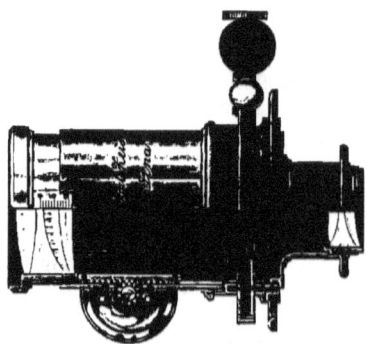

FIG. 57.

number of tables applied to particular constructions manufactured by the firm.

Dr. Miethe, of Messrs. Voigtlander, Brunswick, adopts the method of employing a portrait combination of high intensity as positive element, added to a triple cemented negative, in his constructions destined for large portraiture in the studio.

There is a theoretical advantage as regards the brilliancy of the image, by keeping the number of reflecting surfaces in any optical combination as small as possible ; and the construction of Messrs. Zeiss fulfils this condition in reducing them to the minimum.

The degree of brilliancy due to a greater or less number of reflect-

PRACTICAL APPLICATIONS OF LENS

ing surfaces is at least difficult to trace in practical photographic work, particularly in the studio.

Soft Effects in large Portraiture.—Since the early sixties a considerable number of photographers, aiming at artistic effects, have sought a softer degree of definition than is given by a perfectly corrected photographic lens. The latter defines too well for their purpose, and, in fact, shows up facial defects which are not even visible to the eye, some, indeed, being beneath the skin. In this case the retoucher's skill must be requisitioned, and upon this, if artistically

FIG. 58.

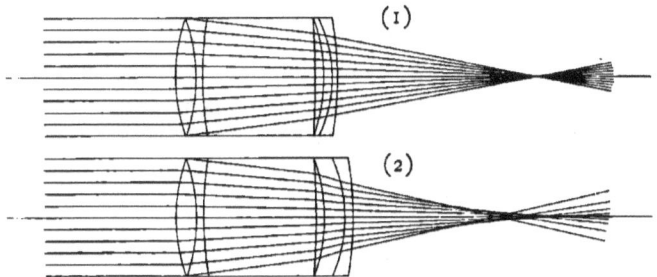

applied, will the true likeness largely depend, even more perhaps than upon the drawing originally given by the lens.

In the portrait lens designed by the late J. H. Dallmeyer, a mechanical means of adjustment was given in order that the image might gradually depart from keen definition to an increasing softness in the image. This device introduces what is technically known as "Spherical Aberration." Fig. 58 roughly indicates how the effect is produced. (1) and (2) represent the posterior portion of the portrait combination. In (1) the lenses are very close together, and the rays leaving the combination converge accurately to a point, the focus. In

TELEPHOTOGRAPHY

(2) the lenses are separated from their normal position by unscrewing the hindermost lens, and the rays no longer converge accurately to one point; those refracted from the margin of the lens cross the axis nearer to the lens than those near the centre. The position where the image of any point in object is now best defined is no longer a point, but a small bundle of rays. Now points in the object situated on either side of the chief point or plane focused upon are represented in (1) by circles of indistinctness which contrast strongly with the finest definition in the chief plane focused for; in (2) by circles of indistinctness which have far less contrast with the definition in the chief plane focused for. The result by (1) may be compared to a drawing in which one plane is drawn by a finely pointed hard pencil and the rest by blunter points of the same pencil: the result by (2) may be compared to a drawing produced throughout by a blunt-pointed soft pencil. So that by sacrificing fine definition in one plane, we produce a softer but more uniform type of definition, which gives the appearance of greater "depth of focus." This softening, due to the introduction of spherical aberration, does not destroy the modelling of the subject, nor detract from the massing of light and shade, but greatly reduces and frequently eliminates the necessity for retouching.

If a portrait lens of this type forms the positive element of a Telephotographic system we can produce a soft primary image and enlarge it by the negative element. The results will seldom require the retoucher's aid, and yet faithfully portray the leading characteristics in the features of the subject. Plate V. was produced in this manner but should not be viewed from too near a standpoint.

In 1893 Mr. J. S. Bergheim carried out some interesting and original experiments by combining single uncorrected positive and negative lenses with a view of suppressing unnecessary detail and aiming at a "quality" in definition he deemed better than that given by corrected lenses. His "Cinderella," for example, shown at the Royal Photographic Society's Exhibition in 1894, showed what his lens was capable of doing in the hands of an artist.

Mr. Bergheim placed the results of his investigations at the author's

PRACTICAL APPLICATIONS OF LENS

disposal for development. Fig. 59 illustrates the lens in its present form, showing the single uncorrected lenses of which it is composed. By placing the diaphragm in *front* of the front lens, distortion is eliminated. In this construction m is very low, and hence it has been possible to make the back single lens of rather larger diameter than the front, enabling a large amount of subject to be included upon the plate when the camera extension is considerable.

Rigidity and considerable variation in the camera extension are essential in a camera for studio work; Fig. 60 illustrates the type of

FIG. 59.

camera and stand suitable for the purpose when large plates are to be covered. The ordinary studio camera, in which there is the usual amount of extension, will answer for the purpose of producing large heads such as those to which reference has been made.

II.—MEDICAL AND SURGICAL PHOTOGRAPHY

Hitherto the application of photography to the sciences of medicine and surgery has been somewhat limited on account of the necessity of bringing the lens very near to the subject in order to obtain images sufficiently large to show the necessary details. When a lens is brought in close proximity to the subject, it has been found that inconveniences,

TELEPHOTOGRAPHY

such as breathing upon the lens, nervousness in the subject, the obstruction of light caused by the instrument itself, the lack of depth of definition, &c., have interfered with an otherwise valuable practice. The increased distance between object and lens rendered possible by use of the Telephotographic combination enables the surgeon to obtain

FIG. 60.

Studio Camera suitably mounted on heavy square table stand, for use in telephotographic portraiture.

records of operations, skin diseases, and various phases of changed conditions in disease, &c., which were formerly practically impossible. The author has only indicated the application to this branch of photography in photographs of the human eye taken under quite ordinary conditions. Plate VI. shows two eyes of children, natural size, and an adult eye enlarged, but taken direct. In each case the

PRACTICAL APPLICATIONS OF LENS

lens was 4 ft. from the subject. For a reduced scale in the image it is obvious that a considerable distance may intervene between the lens and the subject to be photographed. High intensity in the positive element of the system is of course essential.

III.—GROUPS AND HAND-CAMERA WORK BY MEANS OF THE TELEPHOTOGRAPHIC LENS

It will have been observed from the premises that if we include a foreground subject, such as a figure, under a large angle, we must, in order to obtain a sufficiently large image from a given standpoint, necessarily include a large angle of subject in the receding planes, which become dwarfed and insignificant in value. The further we remove the lens from the foreground of our subject, the smaller is the angle under which it is seen or photographed, and the receding planes have now greater importance than in the former case, although the whole is in a diminished scale. If by the Telephotographic lens we make the subject in the foreground of the same size as before at this greater distance, it is obvious that very much less subject will be included in the receding planes, and that they will be rendered in a larger scale than before, possessing due importance.

One frequently notices in a group of figures produced by ordinary photographic means, the rapid diminution in the scale of the images of persons in the receding planes of the group. This is easily obviated with the Telephotographic combination, by taking up a position in which the individuals forming the group are represented more nearly equal in size, and producing the required scale in the final image by the magnification obtained by the negative element of the system in conjunction with the requisite camera extension.

Ordinary common-sense considerations will enable the photographer readily to grasp the bearing of this subject. If we observe from a distance a moving object, such as a railway train, travelling directly in a line towards us, an advance in our direction of a hundred yards or so seems to make but very little difference in the size of the object itself;

TELEPHOTOGRAPHY

but as it continues to approach, equal increments of distance have a greater and greater effect upon its appearance in perspective drawing until, when it is so near that its distance from us is some small multiple only of its entire length, the perspective in which it is seen alters rapidly. The problem which the photographer has to solve in every case where there is considerable depth in his subject is : how far from the foreground his standpoint should be in order to maintain the perspective rendering in a degree which is not overdone in either of the two directions possible. On the one hand, by loss of perspective in taking the subject from too distant a standpoint, and thus diminishing the due effect of distance; or, on the other, by choosing too near a standpoint and so causing the receding planes of his picture to diminish too rapidly in scale, and thereby unduly extending the effect of distance. For the purpose of picture making, irrespective of truthful representation of the subject, the former error is the less objectionable. It must be noted that the perspective effect can be readily observed in the image formed by the positive lens alone, a small portion of the image near the centre of the plate only being considered. When the perspective effect of this small central portion is found satisfactory, the photographer will then increase its scale by means of the negative element of his Telephotographic system, and thus cause that small portion to fill the entire plate on the enlarged scale.

The Telephotographic lens applied to the hand-camera enables the photographer not only to obtain a larger scale in the image than is possible with a lens of ordinary construction, using the same camera extension, but also to produce the same scale given by the latter at a greater distance, with the accompanying improved perspective rendering. To give an example : suppose we take as our positive element a small portrait combination working at $f/3$ of 5 inches focal length, and combine it with a negative element of $2\frac{1}{2}$ inches focal length, and set the screen at a distance of 5 inches from the negative element, we shall produce an image three times as large as that given by the 5-inch lens alone. In the case of ordinary studies of figure subjects we shall now be able to take up a standpoint three times as far as would be neces-

PRACTICAL APPLICATIONS OF LENS

sary to produce the same size of image with the 5-inch lens. The drawing will be vastly better, even in the foreground of the subject, and the receding planes will have much more importance and will not diminish in size so rapidly, while the amount of subject included in the background will of course be considerably less, and generally contribute to a more pictorial effect. With the lens we have described, the size of the image will correspond to that given by an ordinary lens of 15

FIG. 61.

Telephotographic Hand-camera, which may be used with positive lens alone, by placing dark slide in the opening nearest the lens, or two degrees of magnification may be obtained when using the telephoto attachment by placing the slide in one of the other openings.

inches focal length, and as the positive element works at an initial intensity of $f/3$, and its image has been subjected to a magnification of three times, the combination will work at an intensity of $f/9$, which is sufficiently rapid for the ordinary requirements of "instantaneous" hand-camera work. Many useful developments in this branch of photography suggest themselves. Fig. 61 represents a hand-camera in which the positive element is a stigmatic portrait lens of 6 inches focal length combined with a negative element of 3 inches focal length. The box of the camera admits of a separation of 9 inches between the negative lens and the screen, thus giving a magnification of four times

TELEPHOTOGRAPHY

to the image produced by the positive lens alone. We thus have an image equal in size to that produced by an ordinary lens of 24 inches focal length which works at an intensity of $f/16$. A camera of this kind is a veritable "detective" camera, and is valuable for rendering large images on small plates in studies of animal and bird life, ships at sea, distant mountain scenery, &c., and generally in all cases where the rendering of distance is inadequate with the ordinary hand-camera.

IV. *Telephotography for distant Subjects.*—The application of the Telephotographic lens for photographing distant objects has been that most generally employed. In this connection the Telephotographic lens does not differ in its performance from a lens of ordinary construction of the same focal length, but its advantage over the latter lies in the fact that we may produce very large images corresponding to those given by lenses of very great focal length without the necessity for the accompanying great camera extension which the latter involve. The fact that almost any positive system may be converted into a Telephotographic combination by combining it with a negative element whose function is simply to magnify the primary image has made the practice of Telephotography both wide-spread and popular.

It is preferable that the positive element composing the system should have at any rate an intensity not less than $f/8$ for the reason given in Chapter VI., viz., that the limit of magnification which may be given to the image produced by it is comparatively small, without introducing the deleterious effect of diffraction. This, as we have said, commences to assert itself when the intensity of the compound system is below $f/71$.

It is obvious, then, that the effective aperture of the positive element controls the limit of magnification, corresponding to a certain equivalent focal length, which must not be exceeded. The equivalent lens must not be more than 71 times the effective aperture of the positive system; so that if very high magnification is desired, our positive element must be large in diameter. A positive lens of 1-inch aperture must never be converted into a Telephotographic system in

PRACTICAL APPLICATIONS OF LENS

which an equivalent focal length of more than 6 ft. should be brought about; with a 2-inch positive lens, 12 ft., and so on.

The particular constructions which have been designed for the purpose include as positive elements: portrait lenses, any non-distorting doublet, such as a rapid rectilinear, the antiplanat, anastigmatic, stigmatic, &c. We will now shortly describe the various Telephotographic constructions which are in use.

The author's original construction (1891) consisted of a single cemented positive element of high intensity combined with a single

FIG. 62.

Stigmatic lens $\frac{F}{6}$ converted into Telephotographic lens of moderate power by the addition of a negative lens of $\frac{1}{2}$ its focal length, $\frac{f_1}{f_2} = m = 2$.

cemented negative element, with a diaphragm placed between the combinations. The inherent defect of pincushion distortion led him to abandon this form of combination. The chief aim, when the Telephotographic construction was first made, was to dwell on the astounding results produced by high magnification; and, as already pointed out, if the ratio of the focal lengths of the elements composing the system, or m, is high, the distortion is so palpable as to render its employment highly undesirable. He therefore substituted either a portrait combination of high intensity, $f/3$, or any other non-distorting combination as positive element, of an intensity not less than $f/8$, com-

TELEPHOTOGRAPHY

bining it with a double combination negative element as illustrated in Fig. 62. This form of negative element was constructed to give the greatest freedom from distortion, and was found preferable to the single cemented (triple) combination formerly employed. By employing the portrait lens already referred to, in which the spherical correction is adjustable, the author was enabled to produce a Telephotographic system which is perfectly corrected or able to produce sharp images of either near or distant objects. When the positive element of the system is combined with a negative lens for the purpose of attaining high magnification, the spherical correction cannot be perfect for both

FIG. 62A.

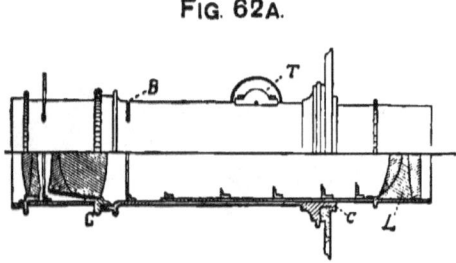

near and distant objects, unless some means be given to regulate the spherical correction. When this is not possible, it is obvious that the only means of attaining a high degree of definition is by use of the diaphragm, involving a loss in rapidity. In the author's combination, the system is corrected to give the finest definition on a moderately near object ; but if it be employed for photographing a very distant object, a slight unscrewing of the back element of the portrait combination will adjust the spherical correction. The necessity for this is most apparent when a high power is employed. The stigmatic lens of intensity $f/4$ also has this means of adjustment, rendering it eminently suitable for the positive element of a Telephotographic system.

PRACTICAL APPLICATIONS OF LENS

Messrs. Steinheil of Munich have given considerable attention to various types of instrument in which the negative enlarging lens is employed. They have produced interesting telescopes of great compactness and yet possessing high power, the magnification being arrived at by the employment of a negative enlarging system which increases the focal length of the object glass without necessitating a long telescope tube. Messrs. Steinheil also construct telescopes for

FIG. 63.

photographic purposes on the same principle, which are designed for high magnification, but to include only a very small angle. For ordinary Telephotographic work, they combine their Group Antiplanat as positive element with a triple cemented negative as shown in Fig. 62A.

Messrs. Zeiss of Jena construct two types of a Telephotographic lens. We have already referred to the form designed for portraiture illustrated in Fig. 57. They recommend their double anastigmat combined with a triple cemented negative lens for all cases in which distortion is inadmissible. This combination is illustrated in Fig. 63.

It will be observed that the same negative lens is used in both

TELEPHOTOGRAPHY

cases; if combined with the single cemented positive lens the curved surface faces the ground glass, but when used in conjunction with a double combination the flat surface is directed to the ground glass. The mounting of this lens has an engraved scale in millimetres which can be read off showing the interval (d of the formula) from which the equivalent focal length of the system can be calculated for any given separation. It is obtained by multiplying the focal lengths of the component lenses together and dividing by the interval which can be read off on the one hand, or set to produce a desired focal length on the other.

FIG. 64.

Dr. Miethe in his Telephotographic construction for all ordinary purposes in which distortion is inadmissible, employs a Collinear lens as the positive element, combining it with a triple cemented negative lens as illustrated in Fig. 64.

We will now call attention to some of the many other applications of the Telephotographic lens in distant photography.

(1) In astronomical work the method of direct enlargement by a negative lens corrected for photographic purposes has of late years superseded the former method of secondary magnification by means of a second positive lens. Plates VII. and VIII. illustrate the application of the negative lens to large telescopes, increasing their focal length in order to produce a larger direct image. In solar photography

PRACTICAL APPLICATIONS OF LENS

generally, the exposures are so short that good results may be obtained with an ordinary camera which is not mounted equatorially, and valuable records of important sun spots and partial or total eclipses of the sun may be readily obtained in ordinary photographic practice. The diameter of the image given of the sun or moon is approximately $\frac{1}{10}$ of an inch for every 10 inches of focal length, and it is readily seen that almost any Telephotographic system will enable us to produce an image of sufficient size to be of interest.

In lunar photography, however, it is not possible to attain a very high degree of definition, owing to the apparent complex movement of the moon during the period of exposure which is necessary for obtaining a good negative; but, nevertheless, records of partial and total eclipses of the moon in progress are sufficiently well defined to be in themselves interesting. A few years ago the author exhibited some lunar photographs at the Camera Club, which have been referred to as warranting further efforts in this direction.

(2) The value of the instrument in photographing distant mountain scenery is already well known and much practised. The author has reproduced as the frontispiece a reduced copy of the historical photograph by M. Boissonas of Geneva, taken at a distance of 44 miles direct in a camera of 5 ft. extension on a 20 × 16 plate. In order that the same size of image could be produced by a lens of ordinary construction a camera 25 ft. in length would have had to be employed. Plates IX. and X. show more modest results with a quarter-plate camera, which can be carried by any enthusiastic mountain climber.

(3) Plates XI. and XII. exhibit the application to long-distance photography generally, and are purposely included as examples of what can be accomplished across the water, even in this climate, where atmospheric conditions are seldom favourable. This leads us to say that the application of the Telephotographic lens in distant photography, even when atmospheric conditions are unfavourable, enables us frequently to obtain results which are far more distinct than the impression given to the eye.

(4) The geologist will appreciate the results shown in the compara-

TELEPHOTOGRAPHY

tive work illustrated in Plates XIII. and XIV. In this case the cliff is at a distance of more than five hundred yards from the camera, and the details shown will point to the application of the Telephotographic lens to studies of geological formations which are inaccessible.

(5) The application of the lens to balloon photography is illustrated in Plate XV., XVI., and XVII. In this application it is needless to say that the positive element of the combination must have a very high intensity. The instrument designed by the author for this purpose consists of a portrait combination of large diameter and high intensity combined with a negative lens which is also of large diameter in relation to its focal length. The value of this instrument to military departments is evident.

(6) Plate XVIII. illustrates the possibility of obtaining naval records in warfare, and suggests the use of the instrument for becoming fully acquainted with details of foreign naval equipments even in times of peace.

(7) The architect and student of archæology will find a wide field in the application of the lens in recalling details of ancient and historical buildings. Portions of carving and sculpture in quite inaccessible positions may be photographed on a large scale on the one hand; or, on the other, large buildings may be photographed from such a distance that they will be rendered without perspective effect, showing the true relative proportions, and being practically a plan in elevation. This branch of the subject has been most clearly exemplified by the work of Mr. E. Marriage, who gained the medal of the Royal Photographic Society in 1895. Fine specimens of the work on a large scale, kindly lent by Mr. Cruickshank, are illustrated in Plates XIX., XX., XXI., and XXII. In this particular branch of work special arrangements in the camera and stand are almost essential. We illustrate in Fig. 65 an arrangement for tilting the camera recommended by Mr. Marriage, and also a strut connecting the tripod and camera for the purpose of greater steadiness. Although the camera may have to be pointed upwards at a considerable angle, the camera back must always be plumb. In these subjects where the duration of exposure is not of great moment, we can afford to employ almost the limit diaphragm for the

PRACTICAL APPLICATIONS OF LENS

combination, and in this case the pencils which form the image are so

FIG. 65.

attenuated that it is possible to obtain fine definition even when the

TELEPHOTOGRAPHY

camera back has been considerably swung to obtain the true plumb.

In Fig. 65 we illustrate a useful device for employment when photographing interiors of cathedrals, churches, &c., in order to prevent the legs of the camera from slipping during exposure. The underside of the three arms is lined with rubber, the three arms themselves being made of soft wood to support the points of the tripod legs.

(8) Plates XXIII., XXIV., and XXV. serve to illustrate a branch of photography referred to in the previous section, as also many similar conditions. The whale illustrated in Plate XXIII. can only be represented by an ordinary lens in painful fore-shortened perspective upon the plate, as, owing to the cramped conditions, it is impossible to obtain a broadside image of it from a view-point on the pier, or one which properly shows the relative proportions of the animal. To obtain these it was necessary to take the camera along the shore to a position where a broadside view could be obtained, necessitating a distance of upwards of three hundred yards from the whale itself. Plate XXIV. shows the result obtained by the same lens that was employed for the near view. The size of the image is now too insignificant to be of value in itself, or to bear enlargement from the photographic negative without loss of detail. By employing a negative lens in conjunction with the same positive element, we produce the result shown in Plate XXV. from the same standpoint.

(9) The application of the Telephotographic lens to animal and bird life has already been roughly hinted at. For some years Mr. R. B. Lodge of Enfield has devoted much patient care to this subject. His methods have been published in several photographic periodicals and he has lectured on the subject before the Royal Photographic Society, who awarded him a medal in 1895 for his work in this branch. Plate XXVI. is indicative of results that may be obtained in this field of work.

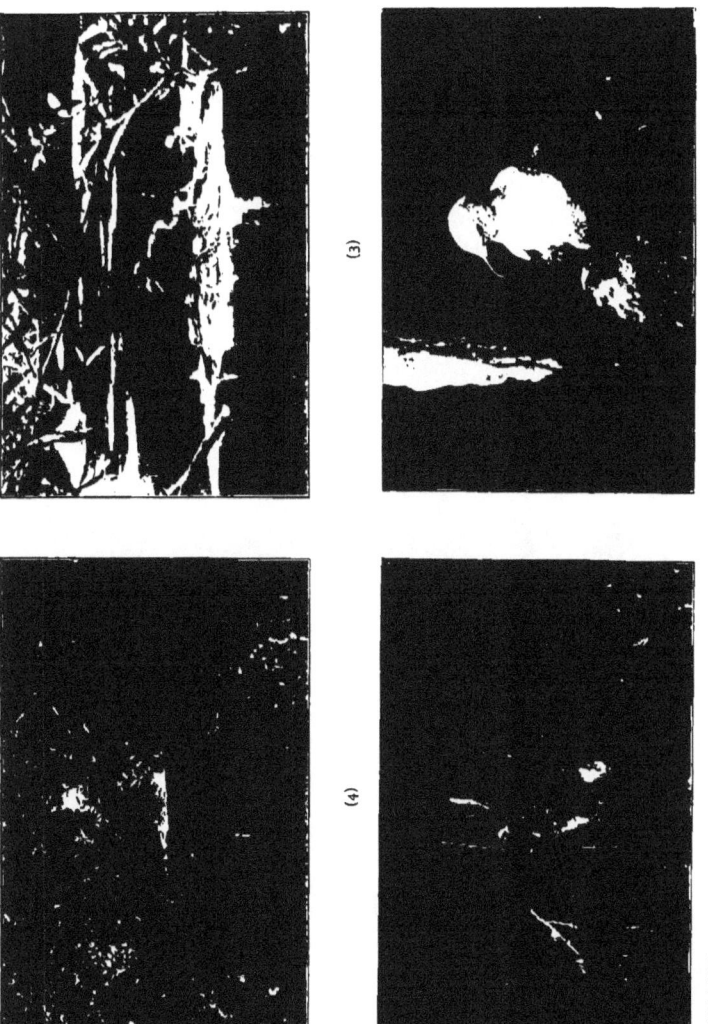

PLATE XXVI

(1) **Little Grebe on Nest.** Taken with Rapid Rectilinear lens 10½ in. focal length. Distance 20 yards.
(2) **Little Grebe on Nest.** Taken with Telephotographic lens from same standpoint. Exposure 2 seconds.
(3) **Kittiwake on nest.** Taken with Telephotographic lens. Distance 20 ft. Exposure ½ second.
(4) **Rabbit.** Taken with Telephoto lens. Distance 20 ft. Exposure ½ second.
(These pictures are the copyright of, and were kindly lent by, Mr. R. B. Lodge of Enfield.)

CHAPTER VIII

WORKING DATA

THE camera and stand must be perfectly rigid. A large tripod head and a bigger clamping screw than usual are advantageous. When long extensions of camera are employed, either a strut attached to the tripod, or an independent support should be given to the back end of the camera to prevent vibration during exposure. The importance of guarding against vibration cannot be insisted upon too strongly. Any slight movement of the camera corresponds to a displacement of the short end of a lever, and the definition is impaired by a corresponding movement of the long end.

When it is possible to choose atmospheric conditions, the clearer and quieter the atmosphere the better. The clearest *appearance* of the atmosphere is not always to be relied upon in photographing distant objects; tremor due to differences in the temperature of air currents, evaporation of moisture, &c., may frequently interfere with the definition given. If these effects are not visible to the eye, a careful examination of the image on the ground glass may reveal them, showing a slight appearance of "dancing." These drawbacks are most likely to occur in hot weather, when the atmosphere is seldom homogeneous.

Focusing must be performed with the greatest care and patience. When a considerable magnification is given, focusing is best carried out by means of the rack and pinion on the lens mount, which adjusts the separation of the elements of the system. A very slight movement of this will rapidly throw the image in or out of focus, and it is best to use a magnifier upon the ground glass. Focusing should always be

TELEPHOTOGRAPHY

performed with the largest aperture in the positive lens consistent with good definition, even if a smaller stop is finally employed.

The greater the magnification given, the less crisp in itself will the image appear on the ground glass, as compared to images given by ordinary positive lenses, but details which are invisible in the smaller image are now clearly visible in the magnified image, and can be photographed. The separate bricks, or blocks of stone, for example, in a building may each not be visible in a small image given by a lens of short focal length, yet when the primary magnification has taken place, we are able to obtain an image of sufficient size to render this separation possible. It is evident that calculation of the requisite magnification (or the equivalent focal length of the system necessary) is always possible when we know the size of certain details in the object, and its distance, in order to render the necessary details visible in the image. The scale of the image is determined by the focal length of the lens, and if we ascribe a limit to the actual size of the image for distinctness, the calculation is simple. As a numerical example, say a church clock 6 ft. in diameter is 500 yards distant. To determine the focal length of the lens necessary to give an image $\frac{1}{4}$ of an inch in diameter, the scale is $\frac{\frac{1}{4}}{72}$ or $\frac{1}{288}$; so that the focal length of the combination must be $\frac{500}{288}$ yards, or roughly, 60 inches. Hence, if our positive element has a focal length of 10 inches, the image must be magnified 6 times by the negative element, or the screen must be placed at a distance of 5 times the focal length of the negative lens from it.

A focusing cloth of sufficient size should be employed to allow the corners of the plate to be examined, after the stop with which the photograph is to be taken has been inserted. If the corners are not fully illuminated, a greater extension of camera must be employed, and the operation of focusing repeated. When considerable extension of camera is employed, it may be found necessary to use a "Hooks' Joint" handle in order to focus by an adjustment of the separation between the lenses.

WORKING DATA

If the positive lens has a high intensity, and only moderate magnification is given, the position of sharp focus may be roughly arrived at by means of the rack and pinion as before, the final focusing being carried out by the camera rack, which will now act as a "fine adjustment" in determining the plane of the sharpest focus. In the previous case the pencils will probably be too attenuated to permit of this method

FIG. 66.

Focusing by means of Hook's universal joint; the handle may be made of any length.

of focusing, as the screen may be moved a considerable distance without much alteration in the definition being traceable.

In focusing, attention must be paid to the varying "curvature of field" caused at different camera extensions; the plane of the image being adjusted to equalise any difference between the planes of best definition for centre and edge of plate, when uniformity throughout is desired. If emphasis is required for any particular plane in the object, naturally the sharpest possible definition is given to this plane.

TELEPHOTOGRAPHY

Where orthochromatic methods are advantageous in ordinary photography, they will be pronouncedly so in Telephotography. The utility of proper light filters (yellow screens, &c.) in the lens used in conjunction with orthochromatic plates is of the highest importance for distant work. When an object is very far off and brightly illuminated, the exposure will be found to be considerably shorter than is necessary for a comparatively near object :—The author has found that the increase of exposure presumed to be necessary when using an orthochromatic screen (when the theoretical increase is not more than 2 to $2\frac{1}{2}$ times), may be ignored ; that is to say, normal exposure, as indicated by the exposure meter without the screen, may be given.

Generally speaking, shorter exposures are necessary in photographing distant objects than would be expected; while longer exposures are necessary with the Telephotographic lens than for lenses of ordinary construction *of the same focal length*, when near objects are photographed, the increase of exposure being proportional to the square of the distance between lens and object.

The chief difficulty in distant Telephotographic work is the adequate rendering of contrast in the negative. At great distances we have not only to contend with the lack of homogeneity in the atmosphere already referred to, but with dust and other impurities in the air which reflect light, and contribute to a general haziness. Backed plates should always be used and slow emulsions are to be preferred.

It is advisable to develop slowly (the plate being carefully protected from light), adding the accelerator very gradually, but development should be carried a little farther than is necessary for a negative taken under ordinary conditions. As the tendency in using all Telephotographic lenses is towards inequality of illumination, or a falling off in the intensity of the image towards the edges of the plate, it is always advisable after first flooding the plate, to develop from the edges towards the centre, by allowing the developer to continually move round the edges in a circular manner in the developing dish.

The requirement of emphasising contrasts will suggest many kinds of developer to the photographer. The following formula given by

PLATE XX

Wheel Window, San Zeno, from same point of view as in Plate XIX., taken with 6-in. Ross Universal Symmetrical with 3-in. negative attachment, 15 in. from back of lens to plate, stopped down to $\frac{f}{22}$. 38 seconds exposure.
(*Copyright of and kindly lent by J. W. Cruickshank, Esq.*)

WORKING DATA

Mr. Marriage has been found to produce admirable results in architectural work:

No. 1.—Pyrogallic acid	1	ounce
Sodium sulphite	3	,,
Citric acid	$\frac{1}{4}$	,,
Water to make	10	,,
No. 2.—Washing soda	8	,,
Sodium sulphite	10	,,
Water to make	80	,,
No. 3.—Bromide of potassium	1	,,
Water to make	10	,,
No. 4.—Carbonate of ammonia	1	,,
Water to make	10	,,

For use take 30 minims No. 1, $\frac{1}{2}$ ounce No. 2, and 10 minims No. 3, make up to one ounce with water. If the plate is over exposed, add equal parts of 3 and 4 to the developer (say 15 minims of each per ounce of developer).

Comparative results by an ordinary and a Telephotographic lens of all ancient and interesting buildings, landmarks, &c., printed in a permanent process, stating the standpoint from which the photographs are taken, date, &c., should be sent to Sir Benjamin Stone, M.P., for the National Photographic Record collection. The ordinary photograph will show the intervening country, or buildings as the case may be, when taken from a distance, between the standpoint and the chief object of interest. These records in pairs will have an added interest in years to come.

TELEPHOTOGRAPHY

ABRIDGED FORMULÆ FOR REFERENCE

POSITIVE LENS

If f represents the focal length of an ordinary positive lens, nf any multiple of it,
By the law of conjugate foci :

$$f^2 = xy \quad . \qquad \qquad . \; (1)$$

where $x = nf$, and $y = \frac{1}{n}f$.

If o be the distance of any object from the lens, and i the distance of the image from it :

$$\left. \begin{array}{l} o = (n+1)f \\ i = (\frac{n+1}{n})f \end{array} \right\} \quad \cdots \cdots (2)$$

TELEPHOTOGRAPHIC LENS (A)

If F represents the focal length of the combination, f_1 and f_2 the focal lengths of the positive and negative lenses composing it, d an interval of separation greater than the difference of their focal lengths, and a their entire separation :

$$\text{F} = \frac{f_1 f_2}{d} \quad . \qquad \qquad (3)$$

and the back focal length

$$\text{B F} = \frac{f_2(f_1 - a)}{d} \quad \cdots \cdots \cdots (4)$$

Calling m the ratio of the focal lengths of positive and negative elements composing the system, or its power :

$$\left. \begin{array}{l} m = \frac{f_1}{f_2} \\ f_1 = m f_2 \end{array} \right\} \quad \cdots \qquad . \; (5) \\ \qquad \qquad \cdots \cdots \cdots . \; (6)$$

or

WORKING DATA

If o be the distance of any object from the Telephotographic lens, and I the distance of the image from it:

$$O = n F + m F + f_1 \quad \ldots \quad \ldots \quad (7)$$
$$I = \frac{1}{n} F + \frac{1}{m} F - f_2 \quad \ldots \quad (8)$$

where n, as above, is any multiple of the focal length of the entire system, and m the focal length of the positive element divided by the focal length of the negative element.

To compare the distance between object and lens, and image and lens in the case of an ordinary positive lens, and in the case of the Telephotographic lens; by equations (2) (7) and (8), we see:

$$O = o + \{F(m - 1) + f_1\} \quad \ldots \quad (9)$$
$$I = i - \left\{ F\left(1 - \frac{1}{m}\right) - f_2 \right\} \quad \ldots \quad (10)$$

Calling d the interval of separation between the positive and negative elements greater than the difference of their focal lengths:

Separation of principal points

$$= \frac{1}{d}(f_1 - f_2 + d)^2 \quad \ldots \quad (11)$$

TELEPHOTOGRAPHIC LENS (B)

Calling M the magnification to which the image formed by the positive element f_1 is subjected by the negative element f_2, and E the camera extension from the negative element to the screen:

$$M = \frac{E}{f_2} + 1 \quad \ldots \quad (12)$$
$$E = f_2(M - 1) \quad \ldots \quad (13)$$

If F represents the focal length of the entire system, and $m, \frac{f_1}{f_2}$, as before,

$$F = M f_1 \quad \ldots \quad (14)$$
$$F = m E + f_1 \quad \ldots \quad (15)$$

TELEPHOTOGRAPHY

If t is the correct exposure to give to the positive element f_1, if used alone, and T that for the entire system:

$$T = M^2 t \qquad (16)$$

If $\dfrac{1}{N}$ represent the scale of the image given by the entire system, and $\dfrac{1}{n}$ that given by the positive lens alone

$$\frac{1}{N} = \frac{1}{n} M \qquad (17)$$

To find the focal length of the entire system when a near object is rendered in a given scale of $\dfrac{1}{N}$,

$$F = \frac{m E + f_1}{\dfrac{m}{N} + 1} \qquad (18)$$

It is evident that when N is ∞, this equation is identical with (15).

If we wish to reproduce an object in a given scale $\dfrac{1}{N}$ by the entire system of definite focal length F, f_1 and f_2 being known, and we propose to magnify the primary image M times (for convenience), in order to do so, we see from (17) that $n = M N$, and hence, distance of object from lens

$$= (M N + 1) f_1 \qquad (19)$$

THE DIAPHRAGM

If a represents the effective aperture of the lens, f its focal length, and I its intensity,

$$I = \frac{a}{f} \qquad (20)$$

Calling E the camera extension, f_1 and f_2 the focal lengths of positive

WORKING DATA

and negative lenses, and d_1 and d_2 their diameters respectively, and D the circle of illumination covered by the system:

$$D = \frac{E}{f_2} \left(\frac{d_1 f_2 + d_2 f_1}{f_1 - f_2} \right) \quad \ldots \ldots \ldots (21)$$

Using the same notation as before, and taking $\frac{1}{100}$ of an inch as limit of indistinctness:

Distance beyond which all objects are in focus

$$= 100\, af + f \quad \ldots \ldots \quad \ldots (22)$$

or

$$= 100\, If^2 + f \quad \ldots \ldots \ldots (22a)$$

For a near object accurately focused:

Front depth,

$$= \frac{nf(n+1)}{100\, If + (n+1)} \quad \ldots \quad \ldots \ldots (23)$$

Back depth,

$$= \frac{nf(n+1)}{100\, If - (n+1)}. \quad \ldots \ldots (24)$$

BIBLIOGRAPHY

"Concave Achromatic Glass Lens, &c." *Proceedings of the Royal Society.* Peter Barlow, F.R.S., and G. Dollond, F.R.S., February and May 1834.

"Eclipse du 28 Juillet 1851, Relevée Héliographiquement," par MM. Vaillat et Thompson, avec un objectif sthenallatique de M. Porro. *Compt. Rend.* 1851.

Bulletin Soc. Franc., pp. 114-117, 1856. J. Porro.

"Enlarged Views by One Operation." *British Journal of Photography*, September 19, 1873. Editor, J. T. Taylor (The Use of the Opera Glass).

"A Novel Enlarging Lens" (Opera Glass). By J. T. Taylor, *British Journal Almanac*, 1877.

"Telescopic Photography Without a Telescope." Editorial comment on the author's lens. By J. T. Taylor, in the *British Journal of Photography*, October 16, 1891.

Dr. A. Meithe's independent work set forth in letter to *British Journal of Photography*, October 30, 1891.

"A New Telescopic Photographic Lens." A Paper read at the Camera Club, London, by the Author, December 10, 1891.

"A Compound Telephotographic Lens." A Paper read at the Camera Club by the Author, March 10, 1892.

Description of "A New Form of Telephotographic Lens." By the Author at the Photographic Society, April 30, 1892. ("The Progress Medal" of the now Royal Photographic Society was awarded in 1896.)

"The Telephotographic Lens." An illustrated pamphlet by the Author, August 1892. (Out of print.)

"Traité Encyclopedique de Photographie." Ch. Fabre. *Prem. Supplem.* 1892.

"Ueber Fernphotographie." By A. Steinheil. *Photo. Correspond.* 1892.

TELEPHOTOGRAPHY

"Telephotography." A Paper delivered at the Society of Arts by the Author. Published in the *Journal*, March 3, 1893. (Awarded the Society's Silver Medal.)

"Telephotographic Systems of Moderate Amplification." A pamphlet by the Author, 1893. (Out of print.)

"Teleobjective." Eder's *Jahrbuch*, 1893, p. 348. (Reference is here made to A. Duboscq's lens.)

"Gebrauchsanleitung fur Teleobjective." Von Dr. P. Rudolph (of the firm of Carl Zeiss, Jena), May 1896.

"Geschichte und Theorie des Photographischen Teleobjectivs." Dr. Von Rohr (of the firm of Carl Zeiss, Jena), April 1897.

"Das Fernobjective." Hans Schmidt, Berlin, April 1898.

PRACTICAL APPLICATIONS OF THE LENS

"Stalking with the Camera." A series of illustrated articles in the *Photogram*. By R. B. Lodge. 1897.

"Telephotography Applied to Architecture." By E. Marriage, in the *Photogram*, January 1898.

"Des Services que peut rendre le Tele-Objective au point de vue Pictorial." By M. Demachy, Photo Club de Paris, in November 1898. An abstract appears in the *Amateur Photographer*, January 13, 1899.

"Telephotography." Illustrated. By Dr. E. J. Spitta, F.R.A.S., in *Photography* March 23, 1899.

roduct-compliance

www.ingramcontent.com/pod-product-compliance
Lightning Source LLC
Chambersburg PA
CBHW020915230426
43666CB00008B/1459